Student Solutions Manual

Alexander Grushow
Rider University

to accompany

CHEMISTRY
Structure and Dynamics

Second Edition

James N. Spencer
Franklin and Marshall College

George M. Bodner
Purdue University

Lyman H. Rickard
Millersville University

John Wiley & Sons, Inc.

Cover Photo: Adri Berger/Stone

To order books or for customer service call 1-800-CALL-WILEY (225-5945).

ISBN 0-471-20496-X

Printed in the United States of America

10 9 8 7 6 5 4 3 2 1

Printed and bound by Victor Graphics, Inc.

Table of Contents

Chapter 1
Elements and Compounds

1 - 1 A **chemist** is a scientist who seeks to understand the composition, structure and properties of substances and the reactions by which one substance is converted into another. This is accomplished by performing experimental tests to recognize patterns of behavior among different substances, then developing models which explain these behaviors, and using the models to predict behavior of other substances. These are the goals of **chemistry**.

1 - 3 Experiments are an important part of chemistry. Through the observations made in an experiment, chemists develop a model to explain the results of the experiment. Further experiments are then performed to test the model's ability to predict the properties of other substances.

1 - 5 The **composition** of a substance indicates what makes up the substance. If the composition of the substance is a **pure material**, the composition is made up of one thing only. If the composition of the substance is a **mixture**, then the substance is made up of more than one kind of material.

1 - 7 An **element** is a substance that contains only one kind of atom. An element cannot be decomposed into a simpler substance. Oxygen, zinc, and bromine are examples of elements. A **compound** is a substance composed of more than one kind of atom. A compound has constant composition. Elements in a compound can be separated by chemical processes. Water, rubber, and carbon dioxide are examples of compounds. A mixture is composed of more than one substance and its composition may vary. Substances that make up the mixture can be separated by physical means. Air, a soft drink, and bronze (a mixture of the two metals copper and tin) are examples of mixtures.

1 - 9 (a) diamond: element
(b) brass: mixture
(c) soil: mixture
(d) glass: mixture
(e) cotton: mixture
(f) milk of magnesia: mixture
(g) salt: compound
(h) iron: element
(i) steel: mixture

1 - 11 The formula P_4S_3 indicates that the elements phosphorus and sulfur are combined in the ratio of four to three. Four atoms of P combine with three atoms of S to form one molecular unit of P_4S_3.

1 - 13 (a) Sb (b) Au (c) Fe (d) Hg (e) K (f) Ag (g) Sn (h) W

1 - 15 (a) titanium (b) vanadium (c) chromium (d) manganese (e) iron (f) cobalt (g) nickel (h) copper (i) zinc

1-17 (a) Co represents the element cobalt. CO represents the compound carbon monoxide which consists of two separate kinds of atoms; one each of carbon and oxygen per molecular unit.
(b) Cs represents the element cesium. CS_2 represents the compound carbon disulfide which consists of two separate kinds of atoms; one of carbon and two of sulfur per molecular unit .
(c) Ho represents the element holmium. H_2O represents the compound water which consists of two separate kinds of atoms; two of hydrogen and one of oxygen per molecular unit.
(d) 4P represents four individual phosphorus atoms. P_4 represents a molecular unit composed of four phosphorus atoms linked together.

1-19 That atoms have different weights that they combine together in whole number ratios can be observed by reacting 2 grams of hydrogen gas with 16 grams of oxygen gas to form 18 grams of liquid water releasing a large amount of energy. This also shows that atoms (matter) cannot be created or destroyed. However, when the same 2 grams of hydrogen can be placed with helium gas and no reaction will take place, it shows that helium and oxygen have different chemical properties.

1-21 Dalton's assumptions do say that atoms are indivisible and indestructible, but they do not give any clues as to the structure of the atom.

1-23 Fusion experiments have shown that it is possible to take two small atoms and fuse them together to make a single larger atom. This process however requires extreme temperatures and pressures. However it does indicate that the number of atoms in the universe is not a constant.

1-25 Dalton's assumption that all atoms of the same element are identical is in error. For example, it is possible for atoms of the same element to have a different number of neutrons.

1-27 Electron, relative charge: -1
Proton, relative charge: +1
Neutron, relative charge: 0

1-29 The electron is the particle which has the smallest mass.

1-31 The radius of an atom is approximately 10,000 times larger than the radius of the nucleus.

1-33 24 protons = Chromium (Cr) = 24 atomic number
24 protons + 28 neutrons = 52 mass number
24 protons - 24 electrons = 0 net charge (a neutral atom)
^{52}Cr

1-35 Iodine (I) = 53 protons = 53 atomic number
127 mass number - 53 protons = 74 neutrons
neutral atom so the number of protons and electrons is equal: 53 electrons
Z=53
mass number = 127

1-37 34 protons=Selenium (Se)
34 protons+45 neutrons=79 mass number
34 protons-34 electrons=0 net charge; neutral atom
^{79}Se

1-39

	Z	A	e
^{31}P	15	31	15
^{18}O	8	18	8
^{39}K	19	39	19
^{58}Ni	28	58	28

1-41 The mass of ^{6}Li is 6.01512 amu. The mass of ^{1}H is 1.007825 amu. A ^{6}Li atom is 5.96842 times more massive than a ^{1}H atom.

1-43 $\frac{\text{mass of } ^{12}C}{\text{mass of } X} = 0.750239$, mass of X =15.9949. Atom X is ^{16}O.

1-45

Mass (grams)	Z	A	Number of neutrons	Mass (amu)
1.6627×10^{-23}	5	10	5	10.0129
3.9829×10^{-23}	12	24	12	23.9850
2.9888×10^{-23}	8	18	10	17.9992
1.7752×10^{-22}	47	107	60	106.903

1-47 Three, ^{16}O, ^{17}O, ^{18}O

1-49 1.99265×10^{-23} g; 12.0000 amu

1-51 (b) The random selection will include isotopes ^{12}C and ^{13}C, but it will be mostly ^{12}C.

1-53 H^+ represents a hydrogen atom which has lost an electron to become a positive ion. The single electron of a hydrogen atom moves around the nucleus and makes the apparent size of the atom about 2000 times larger than the H^+ ion. H_2 represents a diatomic molecule in which two hydrogen atoms are chemically bonded to each other.

1-55 Atomic number = number of protons =24, chromium (Cr)
21 electrons which is three less than the number of protons, +3 charge
mass number= 24 protons + 28 neutrons = 52.
$^{52}Cr^{+3}$

1-57 34 protons=Selenium (Se)
34 protons+45 neutrons=79 mass number
34 protons-36 electrons=-2 net charge
$^{79}Se^{-2}$

1-59 Polyatomic ions are electronically charged substances which are composed of more than one atom. Note that a polyatomic ion is a single unit with a singular charge.

1-61 +1: Ammonium, NH_4^+
Hydronium, H_3O^+

1-63 Verdigris, $Cu_3(OH)_2(CH_3CO_2)_4$ contains two OH^- and four $CH_3CO_2^-$ ions, therefore there is a charge of: 6(−1)=−6 which needs to be balanced off by the three copper ions. The copper ions must then each have a charge of +2.

1-65 (a)Na_2O_2 (b)$Zn_3(PO_4)_2$ (c)K_2PtCl_6

1-67 Potassium peroxide, K_2O_2

1-69 For the Fe^{2+} ion, the formula for the oxide is FeO. For the Fe^{3+} ion, the formula is Fe_2O_3. Combining the two, we get Fe_3O_4.

1-71 Compounds of the elements Cu and Ag with oxygen and chlorine, have the same formula as similar compounds of Li and Na. Ag_2O, Cu_2O, Li_2O Na_2O, AgCl, CuCl, LiCl, NaCl.

1-73 (a) IA (b) IVA (c) IIA (d) VIA (e) IIA (f) VIIIA (g) VIIA

1-75 Seven: H, Li, Na, K, Rb, Cs, Fr

1-77 Similar chemical properties are exhibited by elements in the same group. Sets which are in the same group are (a) and (d).

1-79 (a) N, nonmetal (b) Sb, nonmetal (c) Sc, metal (d) Se, nonmetal
(e) Ge, semimetal (f) Sm, metal (g) Sn, metal (h) Sr, metal

1-81 Elements: Group VA: nitrogen, nonmetal
phosphorus, nonmetal
arsenic, semimetal
antimony, semimetal
bismuth, metal
In general, metal-like properties become more pronounced going down the group.

1-83 $10.0\ cm^3 \times 7.9\ \frac{g}{cm} = 79\ g.$

$5.0\ cm^3 \times 10.5\ \frac{g}{cm} = 53\ g.$

The sample of iron weighs more.

1-85 $\frac{271g}{20.0\,mL} = 13.6\ \frac{g}{cm}$

1-87 The ion has 18 electrons, 20 protons, and 20 neutrons. The chemical symbol for X is Ca.

1-89

classification	group	period	electrons	element
metal	IA	3	11	Na
semimetal	IVA	4	32	Ge
semimetal	IIIA	2	5	B
semimetal	IVA	3	14	Si
nonmetal	VIIA	4	35	Br

Chapter 2
The Mole: The Link Between the Macroscopic and the Atomic World of Chemistry

2-1 (b) a gold ring and (d) gold dust are macroscopic scale.

2-3 Solid gold is represented by Au(s).
Fe(s) would symbolize a solid bar of iron.
Fe(g) would symbolize a single iron atom in the gas phase.

2-5 The atomic weight of zinc is 0.486(63.9291 amu) + 0.279(65.9260 amu) + 0.041(66.9721 amu) + 0.188(67.9249 amu) + 0.006(69.9253 amu)= 65.4 amu

2-7 24.305 amu

2-9 Silicon, Si

2-11 (a) The element is lithium, Li. The average atomic mass of Li is 6.941 amu, just a little less than 7.01600 amu the mass of the predominant isotope of element X.
(b) Both isotopes have 3 protons and 3 electrons. The ^{7}Li has 4 neutrons and a mass number of 7. The other isotope is ^{6}Li, which has 3 neutrons and a mass number of 6.
(c) Lithium is a group IA element which forms +1 ions. This is confirmed by the compounds with Br (which forms a –1 ion), SO_4 (which is a –2 polyatomic ion) and PO_4 (which is a –3 polyatomic ion). The Li^+ ion will have two electrons.

2-13 (a) The atomic mass is a weighted average of the relative masses. Bromine has an atomic mass in between the relative masses of the isotopes.
(b) The atomic mass of Br is 79.904 amu almost exactly halfway in between the relative mass of ^{79}Br and ^{81}Br, therefore both isotopes occur at about a 50:50 abundance.

2-15 (c) is the True answer. The average mass of a lithium atom is 6.941 amu. Therefore the mass of 100 selected at random will be 100 times the average mass.

Looking back at Table 1.4, 90.51% of all neon atoms will have a mass number of 20. Therefore of 10,000 random atoms, 9,051 will have a mass number of 20.

2-17 The element is aluminum since the mass of 26.982 is on the periodic table.

2-19 12.011 amu x 4.33 = 52.0 amu The element is chromium.

2-21 (b), (c) an (d)

2-23 Trick question. They are the same. A mole of Fe has 6.022 x 10^{23} atoms, as does a mole of Cu.

2-25 a)Ni b) Zn c) Ge d) Pb

2-27 2:1

2-29 $2.5 \text{ dozen} \times \left(\frac{\$0.90}{1 \text{ dozen}}\right) = \$2.25$

$2.5 \text{ mols C} \times \left(\frac{12.011 \text{ g C}}{1 \text{ mol C}}\right) = 3.0 \times 10^1 \text{ g C}$

2-31 $(16 \text{ g } O_2)\left(\frac{1 \text{ mol } O_2}{31.998 \text{ g } O_2}\right)\left(\frac{2 \text{ mol O}}{1 \text{ mol } O_2}\right)\left(\frac{6.022\times10^{23} \text{ atoms O}}{1 \text{ mol O}}\right) = 6.0\times10^{23} \text{ atoms O}$

$(31 \text{ g } P_4)\left(\frac{1 \text{ mol } P_4}{123.896 \text{ g } P_4}\right)\left(\frac{4 \text{ mol P}}{1 \text{ mol } P_4}\right)\left(\frac{6.022\times10^{23} \text{ atoms P}}{1 \text{ mol P}}\right) = 6.0\times10^{23} \text{ atoms P}$

$(32 \text{ g } S_8)\left(\frac{1 \text{ mol } S_8}{256.528 \text{ g } S_8}\right)\left(\frac{8 \text{ mol S}}{1 \text{ mol } S_8}\right)\left(\frac{6.022\times10^{23} \text{ atoms S}}{1 \text{ mol S}}\right) = 6.0\times10^{23} \text{ atoms S}$

2-33 Mass of ^{1}H: $\frac{1.0079 \text{ amu}}{\text{atom}} \times \frac{1 \text{ g}}{6.022\times10^{23} \text{ amu}} = \frac{1.674\text{x}10^{-24} \text{ g}}{\text{atom}}$

Mass of ^{12}C: $\frac{12.000 \text{ amu}}{\text{atom}} \times \frac{1 \text{ g}}{6.022\times10^{23} \text{ amu}} = \frac{1.993\text{x}10^{-23} \text{ g}}{\text{atom}}$

2-35 6.022 x 10^{23} atoms is one mole. One mole of ^{12}C weighs 12.000 g.

2-37 $25.0 \text{g Sn} \times \frac{1 \text{mol}}{118.71 \text{g}} \times \frac{6.022\text{x}10^{23} \text{ atoms}}{1 \text{mol atoms}} = 1.27\text{x}10^{23} \text{ atoms}$

2-39 $4.56\text{x}10^{23} \text{ Li atoms} \times \frac{1 \text{mol atoms}}{6.022\text{x}10^{23} \text{ atoms}} \times \frac{6.941 \text{g}}{1 \text{mol Li}} = 5.26 \text{g}$

2-41 (a) 6.022 x 10^{23} atoms

(b) $2 \text{mol atoms} \times \frac{6.022\text{x}10^{23} \text{ atoms}}{1 \text{mol atoms}} = 1.2044\text{x}10^{24} \text{ atoms}$

(c) $0.5 \text{mol atoms} \times \frac{6.022\text{x}10^{23} \text{ atoms}}{1 \text{mol atoms}} = 3.011\text{x}10^{23} \text{ atoms}$

(d) $0.10 \text{mol atoms} \times \frac{6.022\text{x}10^{23} \text{ atoms}}{1 \text{mol atoms}} = 6.022\text{x}10^{22} \text{ atoms}$

2-43 $\frac{1.6735\text{x}10^{-24} \text{ g}}{\text{atom}} \times 6.022\text{x}10^{23} \text{ atoms} = 1.0077 \text{ g}$

2-45 12.011 amu + 4 x 1.0079 amu = 16.043 amu. 16.043 g

2-47 (a) False. The molar mass of NH_3 is 17.031 g/mol while that of H_2O is 18.015 g/mol.
(b) True.

$$(48\text{ g }CO_2)\left(\frac{1\text{ mol }CO_2}{44.009\text{ g }CO_2}\right)\left(\frac{1\text{ mol C}}{1\text{ mol }CO_2}\right)\left(\frac{6.022\times10^{23}\text{ atoms C}}{1\text{ mol C}}\right)=6.6\times10^{23}\text{ atoms C}$$

$$(12\text{ g C})\left(\frac{1\text{ mol C}}{12.011\text{ g C}}\right)\left(\frac{6.022\times10^{23}\text{ atoms C}}{1\text{ mol C}}\right)=6.0\times10^{23}\text{ atoms C}$$

(c) False, both contain 6.022 x 10^{23} molecules. There are 6.022 x 10^{23} N atoms in one mole NH_3 and 2 x 6.022 x 10^{23} N atoms in one mole of N_2.
(d) False

$$(100\text{ g Cu})\left(\frac{1\text{ mol Cu}}{63.546\text{ g Cu}}\right)\left(\frac{6.022\times10^{23}\text{ atoms Cu}}{1\text{ mol Cu}}\right)=9.48\times10^{23}\text{ atoms Cu}$$

$$(100\text{ g CuO})\left(\frac{1\text{ mol CuO}}{79.545\text{ g CuO}}\right)\left(\frac{1\text{ mol Cu}}{1\text{ mol CuO}}\right)\left(\frac{6.022\times10^{23}\text{ atoms Cu}}{1\text{ mol Cu}}\right)=7.57\times10^{23}\text{ atoms Cu}$$

(e) True. 100 mol of Ni(s) contain 100 x 6.022 x 10^{23} atoms of Ni. 100 mol of $NiCl_2$ contain 100 x 6.022 x 10^{23} atoms of Ni.

2-49 (c) 0.050 mol glucose

2-51 (a) CH_4, $12.011 + 4(1.0079) = 16.043\ \frac{g}{mol}$

(b) $C_6H_{12}O_6$, $6(12.011) + 12(1.0079) + 6(15.999) = 180.155\ \frac{g}{mol}$

(c) $(CH_3CH_2)_2O$, $4(12.011) + 10(1.0079) + 15.999 = 74.122\ \frac{g}{mol}$

(d) CH_3CSNH_2, $2(12.011) + 5(1.0079) + 32.066 + 14.007 = 75.135\ \frac{g}{mol}$

2-53 (a) $Cr(CO)_6$, $51.996 + 6(12.011) + 6(15.999) = 220.056\ \frac{g}{mol}$

(b) $Fe(NO_3)_3$, $55.847 + 3(14.007) + 9(15.999) = 241.859\ \frac{g}{mol}$

(c) $K_2Cr_2O_7$, $2(39.098) + 2(51.996) + 7(15.999) = 294.181\ \frac{g}{mol}$

(d) $Ca_3(PO_4)_2$, $3(40.078) + 2(30.974) + 8(15.999) = 310.174\ \frac{g}{mol}$

2-55 Molar mass of MSG, $C_5H_8NNaO_4$:
$5(12.011) + 8(1.0079) + 14.007 + 22.990 + 4(15.999) = 169.111\ \frac{g}{mol}$

2-57 Atomic weight Pt= $\left(\frac{159.4\text{ g Pt}}{0.8170\text{ mol Pt}}\right) = 195.1\ \frac{g}{mol}$

2-59 $(0.0582\text{ mol }CCl_4)\left(\frac{153.823\text{ g }CCl_4}{1\text{ mol }CCl_4}\right)=8.95\text{ g }CCl_4$

2-61 (a) CrO: $\%Cr = \frac{51.996}{67.995} \times 100 = 76.470\ \%$

(b) Cr_2O_3: $\%Cr = \frac{2(51.996)}{151.989} \times 100 = 68.421\ \%$

(c) CrO_3: $\%Cr = \frac{51.996}{99.993} \times 100 = 52.000\ \%$

2-63 $C_{14}H_9Cl_5$, Mol Wt = 14(12.011) + 9(1.0079) + 5(35.453) = 354.490 $\frac{g}{mol}$

$\%C = \frac{14(12.011)}{354.490} \times 100 = 47.435\%$

$\%H = \frac{9(1.0079)}{354.490} \times 100 = 2.5589\%$

$\%Cl = \frac{5(35.453)}{354.490} \times 100 = 50.006\ \%$

2-65 (a) $CaCO_3$: $\%Ca = \frac{40.078}{100.086} \times 100 = 40.044\ \%$

(b) $CaSO_4$: $\%Ca = \frac{40.078}{136.140} \times 100 = 29.439\ \%$

(c) $Ca_3(PO_4)_2$: $\%Ca = \frac{3(40.078)}{310.174} \times 100 = 38.763\ \%$

Tablets that contain $CaCO_3$ are most efficient in delivering Ca^{+2} ions to the body.

2-67 $(15.95\ g\ P_4O_{10})\left(\frac{1\ mol\ P_4O_{10}}{283.886\ g\ P_4O_{10}}\right)\left(\frac{4\ mol\ P}{1\ mol\ P_4O_{10}}\right) = 0.2247$ mols P

2-69 (a) $(0.100\ mol\ KMnO_4)\left(\frac{4\ mol\ O}{1\ mol\ KMnO_4}\right)\left(\frac{6.022\times10^{23}\ atoms\ O}{1\ mol\ O}\right)$

$= 2.41\times10^{23}$ atoms O

(b) $(0.25\ mol\ N_2O_5)\left(\frac{5\ mol\ O}{1\ mol\ N_2O_5}\right)\left(\frac{6.022\times10^{23}\ atoms\ O}{1\ mol\ O}\right)$

$= 7.5\times10^{23}$ atoms O

(c) $(0.45\ mol\ C_{16}H_{17}N_2O_5SK)\left(\frac{5\ mol\ O}{1\ mol\ C_{16}H_{17}N_2O_5SK}\right)\left(\frac{6.022\times10^{23}\ atoms\ O}{1\ mol\ O}\right)$

$= 1.4\times10^{24}$ atoms O

2-71 A 100.00 g sample of stannous fluoride contains 24.25 g F and 75.75 g Sn.

mol F = 24.25 g x $\frac{1\ mol\ F}{18.998\ g}$ = 1.276 mol F

mol Sn = 75.75 g x $\frac{1\ mol\ Sn}{118.71\ g}$ = 0.6381 mol Sn

to determine the ratio $\frac{1.276}{0.6381}$ = 2.000 = 2 mol F $\frac{0.6381}{0.6381}$ = 1.000 = 1 mol Sn

Empirical Formula: SnF_2

2-73 A 100.0 g sample of pyrolusite contains 63.2 g Mn and 36.8 g O.

$\# \text{ mol Mn} = 63.2 \text{ g} \times \frac{1 \text{ mol Mn}}{54.938 \text{ g}} = 1.15 \text{ mol Mn}$

$\# \text{ mol O} = 36.8 \text{ g} \times \frac{1 \text{ mol O}}{15.999 \text{ g}} = 2.30 \text{ mol O}$

to determine the ratio $\frac{1.15}{1.15} = 1.00 = 1 \text{ mol Mn}$ $\quad \frac{2.30}{1.15} = 2.00 = 2 \text{ mol O}$

Pyrolusite has the formula MnO_2. Answer (b).

2-75 A 100.00 g sample of chalcopyrite contains 34.59 g Cu, 30.45 g Fe, and 34.96 g S.

$\# \text{ mol Cu} = 34.59 \text{ g Cu} \times \frac{1 \text{ mol Cu}}{63.546 \text{ g}} = 0.5443 \text{ mol Cu}$ $\quad \frac{0.5443}{0.5443} = 1.000 = 1 \text{ mol Cu}$

$\# \text{ mol Fe} = 30.45 \text{ g Fe} \times \frac{1 \text{ mol Fe}}{55.847 \text{ g}} = 0.5452 \text{ mol Fe}$ $\quad \frac{0.5452}{0.5443} = 1.002 = 1 \text{ mol Fe}$

$\# \text{ mol S} = 34.96 \text{ g S} \times \frac{1 \text{ mol S}}{32.066 \text{ g}} = 1.090 \text{ mol S}$ $\quad \frac{1.090}{0.5443} = 2.003 = 2 \text{ mol S}$

The empirical formula of chalcopyrite is $CuFeS_2$.

2-77 A 100.0 g sample of MDMA contains 68.4 g C, 7.8 g H , 7.2 g N, and 16.6 g O.

$\# \text{ mol C} = 68.4 \text{ g C} \times \frac{1 \text{ mol C}}{12.011 \text{ g}} = 5.69 \text{ mol C}$ $\quad \frac{5.69}{0.51} = 11 \text{ mol C}$

$\# \text{ mol H} = 7.8 \text{ g H} \times \frac{1 \text{ mol H}}{1.0079 \text{ g}} = 7.7 \text{ mol H}$ $\quad \frac{7.7}{0.51} = 15 \text{ mol H}$

$\# \text{ mol N} = 7.2 \text{ g N} \times \frac{1 \text{ mol N}}{14.007 \text{ g}} = 0.51 \text{ mol N}$ $\quad \frac{0.51}{0.51} = 1 \text{ mol N}$

$\# \text{ mol O} = 16.6 \text{ g O} \times \frac{1 \text{ mol O}}{15.999 \text{ g}} = 1.04 \text{ mol O}$ $\quad \frac{1.04}{0.51} = 2 \text{ mol O}$

The empirical formula of MDMA is $C_{11}H_{15}NO_2$.

2-79 $\# \text{ mol Cu} = 9.33 \text{ g Cu} \times \frac{1 \text{ mol Cu}}{63.546 \text{ g}} = 0.147 \text{ mol Cu}$ $\quad \frac{0.147}{0.147} = 1 \text{ mo.Cu}$

$\# \text{ mol Cl} = (14.54 - 9.33)\text{g} \times \frac{1 \text{ mol Cl}}{35.453 \text{ g}} = 0.147 \text{ mol Cl}$ $\quad \frac{0.147}{0.147} = 1 \text{ mo.Cl}$

The empirical formula is CuCl.

2-81 A 100.00 g sample of ß-carotene contains 89.49 g C and 10.51 g H.

$\# \text{ mol C} = 89.49 \text{ g C} \times \frac{1 \text{ mol C}}{12.011 \text{ g}} = 7.451 \text{ mol C}$ $\quad \frac{7.451}{7.451} = 1.000 \text{ mol C}$

$\# \text{ mol H} = 10.51 \text{ g H} \times \frac{1 \text{ mol H}}{1.0079 \text{ g}} = 10.43 \text{ mol H}$ $\quad \frac{10.43}{7.451} = 1.400 \text{ mol H}$

$C_1H_{1.400}$

Since an empirical formula contains the simplest whole number ratio of combination, we need to multiply the subscripts by 5; the formula is C_5H_7 which has a mass of 67.110 g/mol. Furthermore, the molar mass is some multiple of the empirical mass. In this case, $\frac{536.89}{67.11} = 8.00$ times the empirical mass. The molecular formula for ß-carotene is $C_{40}H_{56}$.

2-83 A 100.00 g sample of caffeine contains 49.48 g C, 5.19 g H, 28.85 g N, and 16.48 g O.

$\# \text{ mol C} = 49.48 \text{ g C} \times \frac{1 \text{ mol C}}{12.011 \text{ g}} = 4.120 \text{ mol C} \qquad \frac{4.120}{1.030} = 4.000 = 4 \text{ mol C}$

$\# \text{ mol H} = 5.19 \text{ g H} \times \frac{1 \text{ mol H}}{1.0079 \text{ g}} = 5.15 \text{ mol H} \qquad \frac{5.15}{1.030} = 5.00 = 5 \text{ mol H}$

$\# \text{ mol N} = 28.85 \text{ g N} \times \frac{1 \text{ mol N}}{14.007 \text{ g}} = 2.060 \text{ mol N} \qquad \frac{2.060}{1.030} = 2.000 = 2 \text{ mol N}$

$\# \text{ mol O} = 16.48 \text{ g O} \times \frac{1 \text{ mol O}}{15.999 \text{ g}} = 1.030 \text{ mol O} \qquad \frac{1.030}{1.030} = 1.000 = 1 \text{ mol O}$

The empirical formula of caffeine is $C_4H_5N_2O_1$ which has a mass of 97.097 g/mol. Since the molar mass is 194.193 g/mol, the molecular formula is $C_8H_{10}N_4O_2$.

2-85 Molar Mass of Valium, $C_{16}H_{13}ClN_2O$: $284.74 \frac{\text{g}}{\text{mol}}$

Molar Mass of Librium, $C_{16}H_{14}ClN_3O$: $299.76 \frac{\text{g}}{\text{mol}}$

$\% \text{ C in Valium} = \frac{192.18}{284.74} \times 100 = 67.491 \%$

$\% \text{ C in Librium} = \frac{192.18}{299.76} \times 100 = 64.111 \%$

$\% \text{ H in Valium} = \frac{13.103}{284.74} \times 100 = 4.6016 \%$

$\% \text{ H in Librium} = \frac{14.111}{299.76} \times 100 = 4.7074 \%$

The % C would have to be measured to an accuracy of ±1% and the % H to ±0.1%.

2-87 $11.9 \text{ mg } CO_2 \times \frac{1 \text{ g}}{1000 \text{ mg}} \times \frac{1 \text{ mol } CO_2}{44.009 \text{ g}} \times \frac{1 \text{ mol C}}{1 \text{ mol } CO_2} = 2.70 \times 10^{-4} \text{ mol C}$

$3.41 \text{ mg} \times \frac{1 \text{ g}}{1000 \text{ mg}} \times \frac{1 \text{ mol } H_2O}{18.015 \text{ g}} \times \frac{2 \text{ mol H}}{1 \text{ mol } H_2O} = 3.79 \times 10^{-4} \text{ mol H}$

$\# \text{ g C} = 2.70 \times 10^{-4} \text{ mol C} \times \frac{12.011 \text{ g}}{1 \text{ mol C}} = 3.24 \times 10^{-3} \text{ g C}$

$\# \text{ g H} = 3.79 \times 10^{-4} \text{ mol H} \times \frac{1.0079}{1 \text{ mol H}} = 3.82 \times 10^{-4} \text{ g H}$

Total mg of C and H = 3.62 mg

mg N=4.38-3.62=0.76 mg N

$0.76 \text{ mg N} \times \frac{1 \text{ g}}{1000 \text{ mg}} \times \frac{1 \text{ mol N}}{14.0067 \text{ g}} = 5.4 \times 10^{-5} \text{ mol N}$

Determine # moles of each atom

$\frac{2.70 \times 10^{-4} \text{ mol C}}{5.4 \times 10^{-5} \text{ mol N}} = 5 \text{ mol C} \qquad \frac{3.79 \times 10^{-4} \text{ mol H}}{5.4 \times 10^{-5} \text{ mol N}} = 7 \text{ mol H} \qquad \frac{5.4 \times 10^{-5} \text{ mol N}}{5.4 \times 10^{-5} \text{ mol N}} = 1 \text{ mol N}$

Empirical formula is C_5H_7N, with a mass of 81.1 amu. Since the molar mass of nicotine is 162.235 g/mol, the molecular formula for nicotine is $C_{10}H_{14}N_2$.

2-89 The <u>same</u> number of atoms will be present. The atoms which made up the molecules of liquid gasoline have been changed chemically to gaseous molecules. But the same number are still present, just in a different chemical form.

2-91 Lavoisier observed that the mass of the products of a chemical reaction is the same as the mass of the reactants one starts with.

2-93 Two gaseous hydrogen molecules and one gaseous oxygen molecule can react to form two gaseous water molecules.
This is the same reaction except that the product is liquid water: Two gaseous hydrogen molecules and one gaseous oxygen molecule can react to form two liquid water molecules.

2-95 One gaseous molecule of carbon dioxide and one liquid water molecule can react to form an aqueous molecule of carbonic acid.

2-97 A gaseous molecule of hydrogen and a gaseous molecule of chlorine will react to give two gaseous hydrogen chloride molecules.
When chlorine and hydrogen react, one mole of each will be consumed and two moles of hydrogen chloride are formed.

2-99 (a) $2\text{mol } H_2 \times \frac{2H}{H_2} \times \frac{6.022\text{x}10^{23}\text{ atoms}}{\text{mol}} = 2.4088\text{x}10^{24}\text{ H atoms}$

$1\text{mol } O_2 \times \frac{2O}{O_2} \times \frac{6.022\text{x}10^{23}\text{ atoms}}{\text{mol}} = 1.2044\text{x}10^{24}\text{ O atoms}$

(b) $2\text{mol } H_2 \times \frac{6.022\text{x}10^{23}\text{ molecules}}{\text{mol}} = 1.2044\text{x}10^{24}\ H_2\text{ molecules}$

$1\text{mol } O_2 \times \frac{6.022\text{x}10^{23}\text{ molecules}}{\text{mol}} = 6.022\text{x}10^{23}\ O_2\text{ molecules}$

$$\left(2\text{mol } H_2O \times \frac{2H\text{ atoms}}{H_2O\text{ molecule}} + 2\text{mol } H_2O \times \frac{1\text{ O atom}}{H_2O\text{ molecule}}\right) \times$$

$$\frac{6.022\text{x}10^{23}\text{ atoms}}{\text{mol}} = 3.6132\text{x}10^{24}\text{ atoms}$$

2 moles of product water are formed.

(c) $2\text{mol } H_2O \times \frac{6.022\text{x}10^{23}\text{ molecules}}{\text{mol}} = 1.2044\text{x}10^{24}\ H_2O\text{ molecules}$

2-101 (a) $2\ Pb(NO_3)_2(s) \rightarrow 2\ PbO(s) + 4\ NO_2(g) + O_2(g)$
(b) $NH_4NO_2(s) \rightarrow N_2(g) + 2\ H_2O(g)$
(c) $(NH_4)_2Cr_2O_7(s) \rightarrow Cr_2O_3(s) + 4\ H_2O(g) + N_2(g)$

2-103 (a) $PF_3(g) + 3\ H_2O(l) \rightarrow H_3PO_3(aq) + 3\ HF(aq)$
(b) $P_4O_{10}(s) + 6\ H_2O(l) \rightarrow 4\ H_3PO_4(aq)$

2-105 From $5\text{ mol } O_2 \text{ x } \frac{2\text{ mol } CO_2}{1\text{ mol } O_2} = 10\text{ mol } CO_2$

2-107 $12\text{ mol Cu x } \frac{1\text{ mol CuO}}{1\text{ mol Cu}} = 12\text{ mol CuO}$

2-109 $3\ MnO_2(s) \rightarrow Mn_3O_4(s) + O_2(g)$
$6.75\text{ mol MnO x } \frac{1\text{ mol } O_2}{3\text{ mol } MnO_2} = 2.25\text{ mol } O_2$

2-111 1. Convert grams of CO reacted to moles of CO.
2. Relate moles of CO reacted to moles of CO_2 produced.
3. Convert moles CO_2 to grams.

2-113 $CH_4(g) + 2\ O_2(g) \rightarrow CO_2(g) + 2\ H_2O(g)$

$$10.0 \text{ g } CH_4 \times \frac{1 \text{ mol } CH_4}{16.043 \text{ g}} \times \frac{2 \text{ mol } O_2}{1 \text{ mol } CH_4} \times \frac{31.998 \text{ g } O_2}{1 \text{ mol } O_2} = 39.9 \text{ g } O_2 \text{ consumed.}$$

$$10.0 \text{ g } CH_4 \times \frac{1 \text{ mol } CH_4}{16.043 \text{ g}} \times \frac{1 \text{ mol } CO_2}{1 \text{ mol } CH_4} \times \frac{44.009 \text{ g } CO_2}{1 \text{ mol } CO_2} = 27.4 \text{ g } CO_2 \text{ produced.}$$

2-115 $2\ KClO_3(s) \rightarrow 2\ KCl(s) + 3\ O_2(s)$

$$25.0 \text{ g } KClO_3 \times \frac{1 \text{ mol } KClO_3}{122.548 \text{ g}} \times \frac{3 \text{ mol } O_2}{2 \text{ mol } KClO_3} \times \frac{31.998 \text{ g } O_2}{1 \text{ mol } O_2} = 9.79 \text{ g } O_2$$

2-117 $C_6H_{12}O_6(aq) \rightarrow 2\ C_2H_5OH(aq) + 2\ CO_2(g)$

$$1.00 \text{ kg } C_6H_{12}O_6 \times \frac{1000 \text{ g}}{1 \text{ kg}} \times \frac{1 \text{ mol } C_6H_{12}O_6}{180.155 \text{ g}} \times \frac{2 \text{ mol } C_2H_5OH}{1 \text{ mol } C_6H_{12}O_6} \times \frac{46.068 \text{ g } C_2H_5OH}{1 \text{ mol } C_2H_5OH} \times \frac{1 \text{ kg}}{1000 \text{ g}}$$

$= 0.511 \text{ kg } C_2H_5OH$

2-119 $Ca_3P_2(s) + 6\ H_2O(l) \rightarrow 3\ Ca(OH)_2(aq) + 2\ PH_3(g)$

$$10.0 \text{ g } Ca_3P_2 \times \frac{1 \text{ mol } Ca_3P_2}{182.182 \text{ g}} \times \frac{2 \text{ mol } PH_3}{1 \text{ mol } Ca_3P_2} \times \frac{33.998 \text{ g } PH_3}{1 \text{ mol } PH_3} = 3.73 \text{ g } PH_3$$

2-121 $N_2(g) + 3\ H_2(g) \rightarrow 2\ NH_3(g)$
$4\ NH_3(g) + 5\ O_2(g) \rightarrow 4\ NO(g) + 6\ H_2O(g)$
$2\ NO(g) + O_2 \rightarrow 2\ NO_2(g)$
$3\ NO_2(g) + H_2O(l) \rightarrow 2\ HNO_3(aq) + NO(g)$

$$150 \text{ g } HNO_3 \times \frac{1 \text{ mol } HNO_3}{63.012 \text{ g}} \times \frac{3 \text{ mol } NO_2}{2 \text{ mol } HNO_3} \times \frac{2 \text{ mol NO}}{2 \text{ mol } NO_2} \times \frac{4 \text{ mol } NH_3}{4 \text{ mol NO}} \times \frac{1 \text{ mol } N_2}{2 \text{ mol } NH_3} \times$$

$$\frac{28.014 \text{ g } N_2}{1 \text{ mol } N_2} = 50.0 \text{ g } N_2$$

2-123 $4\ P_4(s) + 5\ S_8(s) \rightarrow 4\ P_4S_{10}(s)$

$$0.500 \text{ mol } P_4 \times \frac{4 \text{ mol } P_4S_{10}}{4 \text{ mol } P_4} = 0.500 \text{ mol } P_4S_{10}$$

$$0.500\ S_8 \times \frac{4 \text{ mol } P_4S_{10}}{5 \text{ mol } S_8} = 0.400 \text{ mol } P_4S_{10}$$

Since S_8 produces fewer moles of P_4S_{10}, it is the limiting reagent. If P_4 is doubled, S_8 is still the limiting reagent, and the amount of P_4S_{10} produced would remain unchanged. If S_8 is doubled, then P_4 becomes the limiting reagent and the yield of P_4S_{10} would be 0.500 mol.

2-125 $H_2(g) + Cl_2(g) \rightarrow 2\ HCl(g)$

$$10.0\ g\ H_2 \times \frac{1\ mol\ H_2}{2.0158\ g} \times \frac{2\ mol\ HCl}{1\ mol\ H_2} = 9.92\ mol\ HCl$$

$$10.0\ g\ Cl_2 \times \frac{1\ mol\ Cl_2}{70.906\ g} \times \frac{2\ mol\ HCl}{1\ mol\ Cl_2} = 0.282\ mol\ HCl$$

Cl_2 produces fewer moles of HCl, therefore it is the limiting reagent.

$$0.282\ mol\ HCl \times \frac{36.46 g}{mol} = 10.3\ g\ HCl$$

To increase the amount of HCl produced, the amount of Cl_2 would have to be increased.

2-127 $2\ PF_3(g) + XeF_4(s) \rightarrow 2\ PF_5(g) + Xe(g)$

$$100.0\ g\ PF_3 \times \frac{1\ mol\ PF_3}{87.968\ g} \times \frac{2\ mol\ PF_5}{2\ mol\ PF_3} = 1.137\ mol\ PF_5$$

$$50.0\ g\ XeF_4 \times \frac{1\ mol\ XeF_4}{207.28\ g} \times \frac{2\ mol\ PF_5}{1\ mol\ XeF_4} = 0.482\ mol\ PF_5$$

XeF_4 produces fewer moles of PF_5; therefore, it is the limiting reagent and 0.482 moles of PF_5 would be produced.

2-129 $Fe_2O_3(s) + 2\ Al(s) \rightarrow Al_2O_3(s) + 2\ Fe(l)$

$$150\ g\ Al \times \frac{1\ mol\ Al}{26.982\ g} \times \frac{2\ mol\ Fe}{2\ mol\ Al} = 5.56\ mol\ Fe,$$

$$250\ g\ Fe_2O_3 \times \frac{1\ mol\ Fe_2O_3}{159.691\ g\ Fe_2O_3} \times \frac{2\ mol\ Fe}{1\ mol\ Fe_2O_3} = 3.13\ mol\ Fe,$$

Fe_2O_3 produces fewer moles of Fe; therefore it is the limiting reagent. The amount of Fe produced is

$$3.13\ mol\ Fe \times \frac{55.847\ g}{1\ mol\ Fe} = 175\ g\ Fe.$$

2-131 Only (b) is a homogeneous mixture.

2-133 $27.3\ g\ HCl \times \frac{1\ mol\ HCl}{36.461\ g\ HCl} = 0.749\ mol\ HCl$

$$M = \frac{moles}{liter} = \frac{0.749\ mol}{0.125\ L} = 5.99\ M$$

2-135 $252\ g\ NH_3 \times \frac{1\ mol\ NH_3}{17.031\ g} = 14.8\ mol\ NH_3$

$$M = \frac{moles}{L} = \frac{14.8\ mol}{1\ L} = 14.8\ M$$

2-137 $5.77\ g\ Cl_2 \times \frac{1\ mol\ Cl_2}{70.906\ g} = 0.0814\ mol\ Cl_2$

$$M = \frac{mol}{L} = \frac{0.0814\ mol}{1.00\ L} = 0.0814\ M$$

2-139 1.00 L of water, plus 1.00 mol of K_2CrO_4 may not be 1.00 L of solution. The student probably made more than 1.00 L of solution, so the solution was less than 1.00 M. A 1.00 liter sample of a 1.00 mol solute per liter solution must be prepared by placing the solute in a container calibrated to hold 1.000 liter. Add water to the container (volumetric flask) to dissolve the solute. Thoroughly mix the solution. Continue to add more water (solvent) until the liquid level has been brought to the calibration mark.

2-141 $\frac{20\ ng}{ml} \times \frac{10^{-9}\ g}{ng} \times \frac{1000\ mL}{L} \times \frac{1\ mol}{315\ g} = 6.3 \times 10^{-8}\ M$

2-143 $\frac{2.75 g\ AgNO_3}{0.250 L} \times \frac{1 mol\ AgNO_3}{169.87 g\ AgNO_3} = 0.0648 M\ AgNO_3$. To make it half as concentrated either add half as much KCl or use twice as much water.

2-145 $0.50 \frac{mol}{L} \times 0.500 L = 0.25 mol$

2-147 The 0.25 M solution is more concentration since it is the higher molarity; more moles per liter.

2-149 (a) $\frac{0.275\ AgNO_3}{0.500 L} \times \frac{1 mol\ AgNO_3}{169.87 g\ AgNO_3} = 0.00324 M\ AgNO_3$

(b) $0.00324 \frac{mol}{L}\ AgNO_3 \times \frac{0.0100 L}{0.500 L} = 6.48 x 10^{-5} M\ AgNO_3$

(c) $6.48 x 10^{-5} \frac{mol}{L}\ AgNO_3 \times \frac{0.0100 L}{0.250 L} = 2.60 x 10^{-6} M\ AgNO_3$

2-151 $0.050 \frac{mol}{L}\ CuSO_4 \times \frac{x L}{2 \cdot x L} = 0.025 M\ CuSO_4$

2-153 $0.10 \frac{mol}{L} HCl \times 0.250 L \times \frac{1L}{6.0 mol\ HCl} = 0.0042 L$. 4.2 ml of 6.0 M HCl would be diluted to 250 ml. The resulting solution will be 0.10 M HCl.

2-155 $1.20 \frac{mol}{L} HF \times 0.100 L = 0.120 mol\ HF$ are present in the initial solution. If you want a final concentration of 0.45 M, what volume must contain the 0.120 mol?

$0.120 mol \times \frac{1L}{0.45 mol} = .267 L$ is the final volume that the 100 mL is diluted to.

2-157 $M_1V_1 = M_2V_2$

(1.00 L)(3.00 M)=(x L)(17.4 M)

$\frac{(1.00\ L)(3.00\ M)}{(17.4\ M)} = 0.172\ L = 172$ ml of 17.4 M acetic acid is needed.

2-159 (0.200 L)(1.25 M)=(x L)(5.94 M)

$\frac{(0.200\ L)(1.25\ M)}{5.94\ M} = x\ L = 0.0421\ L = 42.1\ ml$

Take about 150 ml distilled water and slowly add 42.1 ml of the HNO_3 with mixing. Bring the solution to a final volume of 200 ml with distilled water and mix. The resulting solution is 1.25 M HNO_3

2-161 $2\ NaI(aq) + Hg(NO_3)_2(aq) \rightarrow HgI_2(s) + NaNO_3(aq)$

$$0.045\ L \times \frac{0.10\ mol\ Hg(NO_3)_2}{1\ L} = 4.5 \times 10^{-3}\ mol\ Hg(NO_3)_2$$

$$4.5 \times 10^{-3}\ mol\ Hg(NO_3)_2 \times \frac{2\ mol\ NaI}{1\ mol\ Hg(NO_3)_2} = 9.00 \times 10^{-3}\ mol\ NaI$$

$$\frac{9.00 \times 10^{-3}\ mol\ NaI}{0.25 \frac{mol\ NaI}{L}} = 3.6 \times 10^{-2}\ L \times \frac{1000\ ml}{1\ L} = 36\ ml\ NaI$$

2-163 $H_2C_2O_4(aq) + 2\ NaOH(aq) \rightarrow Na_2C_2O_4(aq) + 2\ H_2O(l)$

$$25.00\ ml\ H_2C_2O_4 \times \frac{0.2043\ mol\ H_2C_2O_4}{1000\ ml} = 5.108 \times 10^{-3}\ mol\ H_2C_2O_4$$

$$5.108 \times 10^{-3}\ mol\ H_2C_2O_4 \times \frac{2\ mol\ NaOH}{1\ mol\ H_2C_2O_4} = 1.022 \times 10^{-2}\ mol\ NaOH$$

$$M = \frac{mol}{L} = \frac{1.022 \times 10^{-2}\ mol\ NaOH}{0.01042\ L} = 0.9808\ M\ NaOH$$

2-165 $C_6H_{12}O_6(aq) + 5\ IO_4^-(aq) \rightarrow 5\ IO_3^-(aq) + 5\ HCO_2H(aq) + H_2CO(aq)$

$$25.0\ ml\ IO_4^- \times \frac{0.750\ mol\ IO_4^-}{1000\ ml} = 0.0188\ mol\ IO_4^-$$

$$0.0188\ mol\ IO_4^- \times \frac{1\ mol\ C_6H_{12}O_6}{5\ mol\ IO_4^-} = 3.75 \times 10^{-3}\ mol\ C_6H_{12}O_6$$

$$M = \frac{mol}{L} = \frac{3.75 \times 10^{-3}\ mol\ C_6H_{12}O_6}{0.0100\ L} = 0.375\ M\ C_6H_{12}O_6$$

2-167 $$74.5\ g\ Cl \times \frac{1\ mol\ Cl}{35.453\ g} \times \frac{1\ mol\ MCl_2}{2\ mol\ Cl} \times \frac{1\ mol\ M}{1\ mol\ MCl_2} = 1.05\ mol\ M$$

$$\text{atomic mass} = \frac{(100.0 - 74.5)\ g\ M}{1.05\ mol\ M} = 24.3\ \frac{g}{mol}$$

The element is Mg.

2-169 The empirical formula of the reactant compound:

$$31.9\text{ g K}\times\frac{1\text{ mol K}}{39.098\text{ g}}=0.816\text{ mol K}$$

$$28.9\text{ g Cl}\times\frac{1\text{ mol Cl}}{35.453\text{ g}}=0.815\text{ mol Cl}$$

$$39.2\text{ g O}\times\frac{1\text{ mol O}}{15.999\text{ g}}=2.45\text{ mol O}$$

divide by smallest:

$$\frac{0.816}{0.815}=1\text{ mol K}\qquad\frac{0.815}{0.815}=1\text{ mol Cl}\qquad\frac{2.45}{0.815}=3\text{ mol O}$$

The empirical formula of the reactant compound is $KClO_3$.

The empirical formula of the product compound is:

$$52.4\text{ g K}\times\frac{1\text{ mol K}}{39.098\text{ g}}=1.34\text{ mol K}$$

$$47.6\text{ g Cl}\times\frac{1\text{ mol Cl}}{35.453\text{ g}}=1.34\text{ mol Cl}$$

Therefore, the product compound is KCl, and the balanced equation for the decomposition is

$$2\ KClO_3(s) \rightarrow 2\ KCl(s) + 3\ O_2(g)$$

2-171
$$1.00\text{ g Cr}\times\frac{1\text{ mol Cr}}{51.996\text{ g}}=0.0192\text{ mol Cr}\qquad\frac{0.0192}{0.0192}=1.00=1\text{ mol Cr}$$

$$0.923\text{ g O}\times\frac{1\text{ mol O}}{15.999\text{ g}}=0.0577\text{ mol O}\qquad\frac{0.0577}{0.0192}=3.00=3\text{ mol O}$$

The formula of the compound is CrO_3

2-173 Cocaine, $C_{17}H_{21}O_4N$, MW=303.36 g/mol

C=67.31%

H=6.98%

N=4.62%

O=21.10%

Aspirin, $C_9H_8O_4$, MW=180.16 g/mol

C=60.00%

H=4.48%

O=35.52%

With a 7.31% difference in the amount of C and a 2.5% difference in the amount of H, it would be possible to distinguish between cocaine and aspirin by elemental analysis of carbon and hydrogen.

2-175 % CO_2 by mass in each of the metal carbonates $= \frac{44.009 \text{ g } CO_2}{\text{MW metal carbonate}} \times 100$

	Carbonate	MW (g/mol)	% CO_2	
(a)	Li_2CO_3	73.89	59.56	
(b)	$MgCO_3$	84.31	52.20	
(c)	$CaCO_3$	100.09	43.97	
(d)	$ZnCO_3$	125.40	35.10	*
(e)	$BaCO_3$	197.34	22.30	

2-177 $3\ Mg(s) + N_2(g) \rightarrow Mg_3N_2(s)$

$Mg_3N_2(s) + 6\ H_2O(l) \rightarrow 3\ Mg(OH)_2(aq) + 2\ NH_3(aq)$

$$15.0 \text{ g } NH_3 \times \frac{1 \text{ mol } NH_3}{17.031 \text{ g}} \times \frac{1 \text{ mol } Mg_3N_2}{2 \text{ mol } NH_3} \times \frac{3 \text{ mol Mg}}{1 \text{ mol } Mg_3N_2} \times \frac{24.305 \text{ g Mg}}{1 \text{ mol Mg}} = 32.1 \text{ g Mg}$$

2-179 $CuS + 2\ H_2SO_4(aq) \rightarrow CuSO_4(aq) + SO_2(g) + 2\ H_2O(l)$

$2\ CuSO_4(aq) + 5\ I^-(aq) \rightarrow 2\ CuI(s) + I_3^-(aq) + 2\ SO_4^{2-}$

$I_3^-(aq) + 2\ S_2O_3^{2-}(aq) \rightarrow 3\ I^-(aq) + S_4O_6^{2-}(aq)$

$$31.5 \text{ ml} \times \frac{1.00 \text{ mol } S_2O_3^{2-}}{1000 \text{ ml}} = 0.0315 \text{ mol } S_2O_3^{2-}$$

$$\% \text{ Cu} = 0.0315 \text{ mol } S_2O_3^{2-} \times \frac{1 \text{ mol } I_3^-}{2 \text{ mol } S_2O_3^{2-}} \times \frac{2 \text{ mol } CuSO_4}{1 \text{ mol } I_3^-} \times \frac{1 \text{ mol Cu}}{1 \text{ mol } CuSO_4}$$

$$\times \frac{63.546 \text{ g Cu}}{1 \text{ mol Cu}} \times \frac{1}{2.50 \text{ g sample}} \times 100 = 80.0\%$$

2-181 $4\ Fe(s) + 3\ O_2(g) \rightarrow 2\ Fe_2O_3(s)$

$3\ Fe(s) + 2\ O_2(g) \rightarrow Fe_3O_4(s)$

Moles of Fe reacted: $167.6 \text{ g Fe} \times \frac{1 \text{ mol Fe}}{55.847 \text{ g}} = 3.001 \text{ mol Fe}$

Grams of O in the iron oxide: 231.6 g - 167.6 g = 64.0 g O

Moles of O in the iron oxide: $64.0 \text{ g O} \times \frac{1 \text{ mol O}}{15.999 \text{ g}} = 4.000 \text{ mol O}$

The oxide formed is Fe_3O_4.

2-183 The intensity of the color red in the final solution depends on how much Br_2 is formed. In each experiment, Br^- is the limiting reagent. For the first experiment, 0.01 mole of Br^- reacts with 0.02 mole of Cl_2. In the second experiment, 0.01 mole of Br^- reacts with 0.05 mole of Cl_2. Since the amount of Br^- is the same in both experiments, the amount of Br_2 formed in each case will be the same. Thus, both solutions will have the same shade of red.

Chapter 3
The Structure of the Atom

3-1 The positive charge is concentrated in small a portion of the atom called the nucleus.

3-3 According to the Rutherford model the nucleus contains the positive charge of the atom and contains most of the mass of the atom. At the time of Rutherford's experiments the distinction between protons and neutrons had not been made.

3-5 Particles have a definite mass and occupy space. Waves have no mass and carry energy as they travel through space. In addition waves have properties of speed, frequency, wavelength, and amplitude.

3-7 For 440 Hz: Assume the speed of sound is 1116 ft/sec

$$\left(\frac{440 \text{ cycles}}{\text{s}}\right) \times \lambda = 1116 \ \frac{\text{ft}}{\text{s}}$$

$$\lambda = \left(1116 \ \frac{\text{ft}}{\text{s}}\right)\left(\frac{1 \text{ s}}{440 \text{ cycles}}\right) = 2.54 \ \frac{\text{ft}}{\text{cycle}}$$

For 880 Hz

$$\left(\frac{880 \text{ cycles}}{\text{s}}\right) \times \lambda = 1116 \ \frac{\text{ft}}{\text{s}}$$

$$\lambda = \left(1116 \ \frac{\text{ft}}{\text{s}}\right)\left(\frac{1 \text{ s}}{880 \text{ cycles}}\right) = 1.27 \ \frac{\text{ft}}{\text{cycle}}$$

As the frequency increased by a factor of 2 the wavelength is decreased by a factor of 1/2. The speed at which the sound travels to your ear remains constant at 1116 ft/s.

3-9 $\nu\lambda = c$

$$\left(5.0 \times 10^{14} \ \frac{\text{cycles}}{\text{s}}\right) \times \lambda = 2.998 \times 10^{8} \ \frac{\text{m}}{\text{s}}$$

$$\lambda = 2.998 \times 10^{8} \ \frac{\text{m}}{\text{s}} \times \left(\frac{\text{s}}{5.0 \times 10^{14} \text{ cycles}}\right) = 6.0 \times 10^{-7} \ \frac{\text{m}}{\text{cycle}}$$

3-11 Red light has a longer wavelength than blue light.

3-13 $25.147 \text{ MHz} \times \left(\frac{1 \times 10^{6} \text{ Hz}}{1 \text{ MHz}}\right) = 2.5147 \times 10^{7} \text{ Hz}$

$\nu\lambda = c$

$$\left(2.5147 \times 10^{7} \ \frac{\text{cycles}}{\text{s}}\right) \times \lambda = 2.998 \times 10^{8} \ \frac{\text{m}}{\text{s}}$$

$$\lambda = \left(2.998 \times 10^{8} \ \frac{\text{m}}{\text{s}}\right)\left(\frac{1 \text{ s}}{2.5147 \times 10^{7} \text{ cycles}}\right) = 11.92 \ \frac{\text{m}}{\text{cycle}}$$

3-15 $6 \text{ nm}\left(\frac{1 \times 10^{-9} \text{ m}}{1 \text{ nm}}\right) = 6 \times 10^{-9} \text{ m}$

$c = \nu\lambda$

$$\nu \times \left(\frac{6 \times 10^{-9} \text{ m}}{\text{cycle}}\right) = 2.998 \times 10^{8} \frac{\text{m}}{\text{s}}$$

$$\nu = 2.998 \times 10^{8} \frac{\text{m}}{\text{s}} \times \left(\frac{\text{cycle}}{6 \times 10^{-9}\text{m}}\right) = 5 \times 10^{16} \frac{\text{cycle}}{\text{s}}$$

3-17 The lines have the following wavelengths, 656.3 nm, 486.1 nm, 434.0 nm, 410.2 nm

656.3 nm: $\nu = 2.998 \times 10^{8} \frac{\text{m}}{\text{s}} \times \left(\frac{1 \text{ cycle}}{6.563 \times 10^{-7} \text{ m}}\right) = 4.568 \times 10^{14} \frac{\text{cycle}}{\text{s}}$

486.1 nm: $\nu = 2.998 \times 10^{8} \frac{\text{m}}{\text{s}} \times \left(\frac{1 \text{ cycle}}{6.563 \times 10^{-7} \text{ m}}\right) = 6.167 \times 10^{14} \frac{\text{cycle}}{\text{s}}$

434.0 nm: $\nu = 2.998 \times 10^{8} \frac{\text{m}}{\text{s}} \times \left(\frac{1 \text{ cycle}}{6.563 \times 10^{-7} \text{ m}}\right) = 6.908 \times 10^{14} \frac{\text{cycle}}{\text{s}}$

410.2 nm: $\nu = 2.998 \times 10^{8} \frac{\text{m}}{\text{s}} \times \left(\frac{1 \text{ cycle}}{6.563 \times 10^{-7} \text{ m}}\right) = 7.309 \times 10^{14} \frac{\text{cycle}}{\text{s}}$

3-19 The color of the light indicates an emission of a discrete amount of energy. This quantity of energy appears to be different for each element. Therefore, the color of light emitted by a particular substance could be used to identify the elements in the substance.

3-21 The shorter the wavelength the greater the energy; yellow.

3-23 $\lambda = 656.3 \text{ nm} = 6.563 \times 10^{-7} \text{ m}$

$$\nu = 2.998 \times 10^{8} \frac{\text{m}}{\text{s}} \times \left(\frac{1 \text{ cycle}}{6.563 \times 10^{-7} \text{ m}}\right) = 4.568 \times 10^{14} \frac{\text{cycle}}{\text{s}}$$

$E = h\nu$

$$E = (6.626 \times 10^{-34} \text{ Js}) \left(4.568 \times 10^{14} \frac{\text{cycles}}{\text{s}}\right) = 3.027 \times 10^{-19} \text{ J}$$

3-25 $\left(\frac{243.4 \text{ kJ}}{\text{mole}}\right)\left(\frac{1000 \text{ J}}{\text{kJ}}\right)\left(\frac{1 \text{ mole}}{6.022 \times 10^{23} \text{ molecules } Cl_2}\right) = \frac{4.042 \times 10^{-19} \text{J}}{\text{molecule } Cl_2}$

$E = h\nu = hc/\lambda$

$\lambda = hc/E$

$$\lambda = \left(\frac{6.626 \times 10^{-34} \text{Js} \times 2.998 \times 10^{8} \text{m/s}}{4.042 \times 10^{-19} \text{J/molecule } Cl_2}\right) = 4.914 \times 10^{-7} \text{m}$$

This radiation falls in the visible light portion of the electromagnetic spectrum.

3-27 The electron in the hydrogen atom may only be found at specific energy levels, i.e. the electron's energy is quantized. For the electron to move from one level to another it must absorb or emit a discrete (quantized) amount of energy, which is the exact difference between the initial and final quantized energy states.

3-29 The Rutherford model of the atom does not specify where electrons might be found in the atom. The Bohr model does.

3-31 As the distance between the electron and nucleus increases the force holding them together decreases according to Coulomb's law.

3-33 According to the Bohr model, the hydrogen atom consists of a positively charged nucleus and an electron orbiting the nucleus at a discrete radius.

3-35 1312 kJ/mol

3-37 From Fig. 3.5 the difference in energy between n=2 and n=3 is

$$(-145.8 \text{ kJ/mol})-(-328.0 \text{ kJ/mol})=182.2 \text{ kJ/mol}$$

$$182.2\frac{\text{kJ}}{\text{mol}}\times 1000\frac{\text{J}}{\text{kJ}}\times\frac{1 \text{ mole}}{6.022\text{x}10^{23}\text{ atoms}}=3.025 \text{ x } 10^{-19}\text{ J}$$

From Fig. 3.5 the difference in energy between n=2 and n=4 is

$$(-82.0 \text{ kJ/mol})-(-328.0 \text{ kJ/mol})=246 \text{ kJ/mol}$$

$$246.0\frac{\text{kJ}}{\text{mol}}\times 1000\frac{\text{J}}{\text{kJ}}\times\frac{1 \text{ mole}}{6.022\text{x}10^{23}\text{ atoms}}=4.085 \text{ x } 10^{-19}\text{ J}$$

From Fig. 3.5 the difference in energy between n=2 and n=∞ is

$$(0)-(-328.0 \text{ kJ/mol})=328 \text{ kJ/mol}$$

$$328.0\frac{\text{kJ}}{\text{mol}}\times 1000\frac{\text{J}}{\text{kJ}}\times\frac{1 \text{ mole}}{6.022\text{x}10^{23}\text{ atoms}}=5.447 \text{ x } 10^{-19}\text{ J}$$

3-39 It is easier to remove from n=2 because is requires less energy to get to the n=∞ state. Therefore the electron is more stable in the n=1 state, because the energy is even lower than n=2.

3-41 The portion of the Bohr model which accounts for the quantization of electron energies states that electrons reside at discrete distances from the positively charged nucleus. Each distance has a discrete (quantized) energy.

3-43 Energy is needed to overcome the attraction by the positive nucleus for the negatively charged electron.

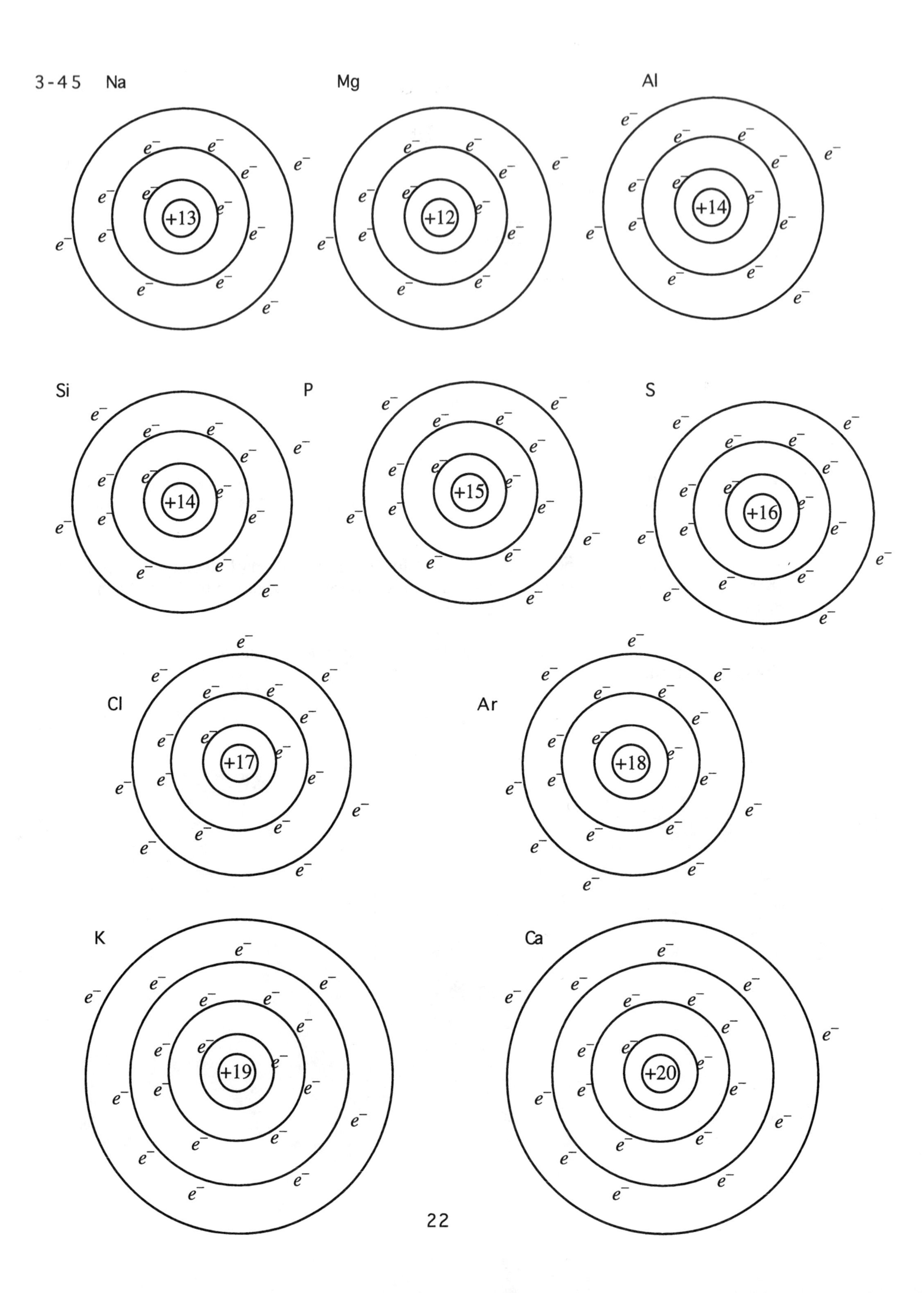
3-45
Na
+13
Mg
+12
Al
+14
Si
+14
P
+15
S
+16
Cl
+17
Ar
+18
K
+19
Ca
+20

3-47 The first ionization energy of F is 1681 kJ/mol and for Cl it is 1251 kJ/mol. Usually after the second row, there are only moderate decreases in first ionization energy as one goes down a column. Given that I is two rows below Cl, one might expect the first ionization energy to be about 900 to 1000 kJ/mol.

3-49 Both Cl^- and Ar have the same number of electrons, however the Cl^- electrons only see a core charge of +7, whereas the electrons in Ar see a core charge of +8. Therefore the Ar will have a higher ionization energy than the Cl^-.

3-51 The electron from the innermost shell is harder to remove because a) it is closer to the nucleus and b) it has no electrons shielding it, so the charge it sees is the full nuclear charge of +3, whereas the outer shell electron only sees a core charge of +1.

3-53 Because the outer shell electrons in chlorine are farther from the nucleus than the outer shell electrons of fluorine.

3-55 The ionization energy of F^- would be less than that of Ne because F^- has one less proton in its nucleus than Ne.

3-57 Both Be and He have the same core charge on their two outer electrons. However, the outer electronic shell of Be is farther away from the nucleus than for He. Therefore, Be will have the lower ionization energy.

3-59 Generally the first ionization energies decrease going from top to bottom of a column of the periodic table.

3-61 The first ionization energy of hydrogen is larger than the first ionization energy of sodium because the $1s^1$ electron of hydrogen is closer to the nucleus and is therefore more tightly held by the nucleus. The core charge on the two atoms is the same.

3-63 Going down a column the core charge felt by the outer electrons is the same, however the size of the outer shell increases. Increasing the distance between the electron and nucleus reduces the coulombic attraction, lowering the ionization energy.

3-65 The element with the smallest first ionization energy is Ca.

3-67 In the first three rows, the core charge, number of valence electrons and group number are all the same.

3-69 No, the photoelectron spectrum of elements in the second row shows that electrons in the second shell will have two different energies after the first two electrons (for Li and Be) are placed in the shell. That is that the valence electrons for B through Ne reside in subshells with two slightly different energies.

3-71 No

3-73 Yes, a core electron can be removed from an atom in a PES experiment if the light striking the atom has a high enough energy to remove a core electron.

3-75 The relative number of electrons which are ionized at that energy.

3-77 The first two electrons of Li are in the 1s subshell, the third electron is in the 2s subshell. This is why there are two peaks. They have different intensities because the relative number of electrons in each subshell is different.

3-79 For a given atom, the lowest energy peak in a PES corresponds to the first ionization energy. Of the atoms shown in Fig 3.14, Li has the lowest first ionization energy.

3-81 Electrons in the 2s and 2p subshells can be removed from B. Only electrons from the 2s subshell will be removed from Be.

3-83 Al

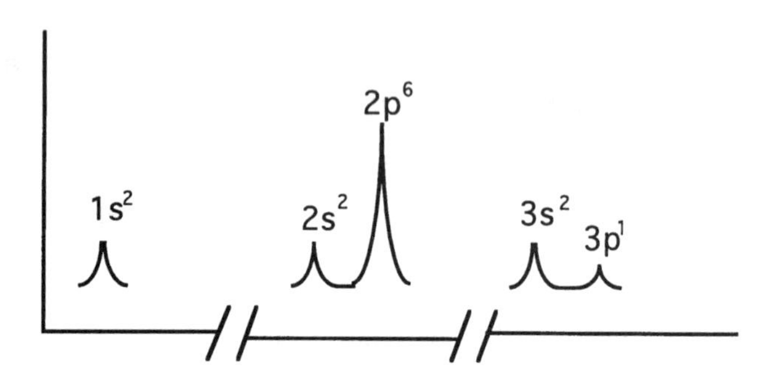

S

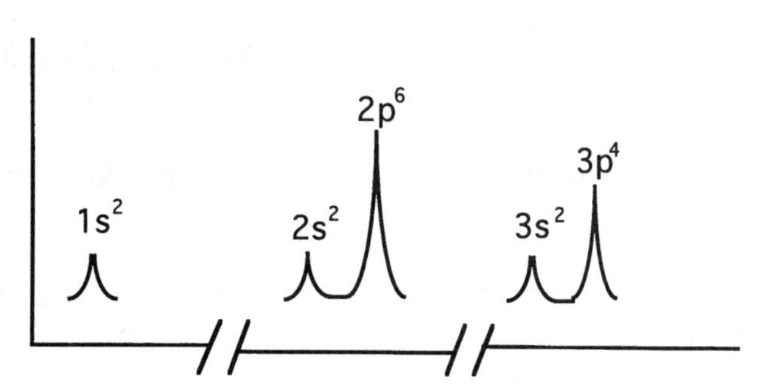

3-85 (a)T he first 18 electrons for K are found in the $1s^2$, $2s^2$, $2p^6$, $3s^2$, and $3p^6$ shells. This is also the order of decreasing ionization energies. The relative intensities for the p electrons would be 3 times as large as the s electrons.

(b) Within a given shell, it is expected that the energies to remove electrons are quite similar. Therefore, if the 19th electron were in an n=4 shell, the ionization energy would be substantially different because the electrons are farther from the nucleus. This can be seen by comparing the relative ionization energies for the different shells in Na and Li. The ionization energy would be closest to 0.42 MJ/mol.

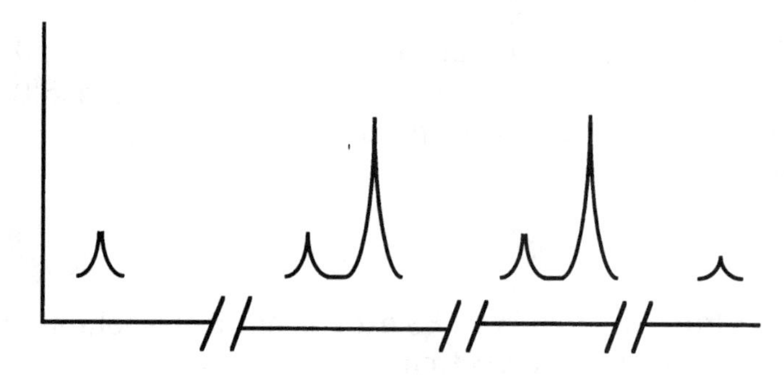

(c) Within a given shell, it is expected that the energies to remove electrons are quite similar. Therefore, if the 19th electron were in an n=3 shell, the energy would be close to but slightly larger than the 1.52 MJ/mole found for Ar. The ionization energy would be closest to 2.0 MJ/mole.

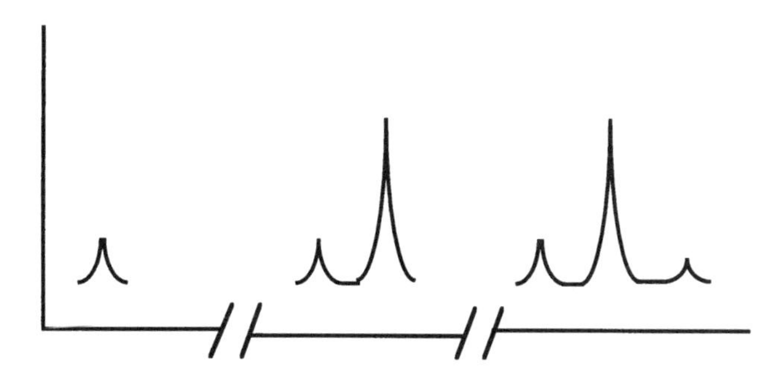

(d) There is a drop in energy from a grouping of 3.93 mJ/mole and 2.38 MJ/mole to 0.42 MJ/mole. This suggests that this final peak represents electrons which are farther from the nucleus and should be assigned to the n=4 shell.

3-87 (a) IIIA (b) +3 (c)+3
(d) Al, the element is in column IIIA and has three distinct electronic shells, putting it in the third period.
(e) λ = 206 nm = 2.06×10^{-7} m

$$\nu = 2.998 \times 10^{8}\ \frac{\text{m}}{\text{s}} \times \left(\frac{1\ \text{cycle}}{6.563 \times 10^{-7}\ \text{m}}\right) = 1.455 \times 10^{15}\ \frac{\text{cycle}}{\text{s}}$$

$E = h\nu$

$$E = (6.626 \times 10^{-34}\ \text{Js})\left(4.568 \times 10^{14}\ \frac{\text{cycles}}{\text{s}}\right) = 0.581\ \text{MJ/mol}$$

3-89 l=0, l=1, l=2,

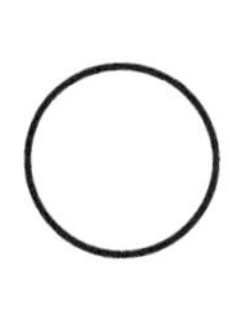

3-91 n = 0,1,2,...,8.
Once n is established then l = 0,..., n-1.
Once l is established then m_l = -l,...,0,..., +l

3-93 If n=4, then l = 0, 1, 2, 3. All orbitals of n=4 are in the same shell. The different l values represent both different angular shapes and subshells.

3-95 If n = 1, l=0, m_l = 0, and m_s = +1/2, -1/2.
If n = 2, l = 0 , m_l = 0, m_s = +1/2, -1/2 and l = 1, m_l = -1,0,+1, and m_s can be +1/2 or -1/2 for each value of m_l.

3-97 9 orbitals in the n=3 shell; 16 orbitals in the n=4 shell; 25 orbitals in the n=5 shell.

3-99 Paired electrons are electrons occupying the same orbital with opposite spin.

3-101 If all the electrons in an atom are spin paired then there can be no net magnetic field.

3-103 Unpaired electrons avoid each other in space.

3-105 (b)

3-107 When n=5 and l=3, m_l = -3, -2, -1, 0, +1, +2, +3.

3-109 The difference between the atomic numbers of any adjacent pair of elements in a group of the periodic table will be 8, 18, or 32 because the change between adjacent pairs corresponds to a filled shell.

3-111 (a)

3-113 There are a total of 16 orbitals that can possess the quantum number n=4. They are a 4s orbital, three 4p orbitals, five 4d orbitals, and seven 4f orbitals. Each of these orbitals can only have one electron with m_s=+1/2. Therefore, there are 16 electrons that can possess the quantum numbers n=4 and m_s = +1/2.

3-115 As n increases the number of subshells increases.

3-117

n	max number of electrons
1	2
2	8
3	18
4	32

3-119 The last electron in Ga is found in a 4 p orbital. Answer (d).

3-121 An orbital is a region of space occupied by a single electron or two electrons of opposite spin.

3-123 To be in the same orbital two electrons must have the same values of n, l, and m_l, but the must have opposite values of spin, m_s.

3-125 (e) is incorrect

3-127 The modern periodic table groups elements such that they exhibit chemical formulas for compounds which are similar to those for other elements associated with the group. The periodic variation in composition is evident by trends along the period.

3-129 The first element expected to have a 3d electron in its electron configuration is in row 4 and column 3 of the periodic table. The element is scandium.

3-131 The element with the electron configuration $[Kr]5s^24d^{10}5p^3$ is found in Group VA. The element is antimony.

3-133 filling of s orbitals groups I and II
filling of p orbitals groups IIIA - VIIIA
filling of d orbitals transition elements
filling of f orbitals inner transition elements

3-135 N [He] $2s^2$ $2p^3$

3-137 x=2

3-139 (c) is correct for the electronic configuration of the Br^- ion.

3-141 6 electrons in s orbitals in the Ti^{2+} ion.

3-143 (d) Y is the first element to have 4d electrons in its electronic configuration.

3-145 (b) satisfies Hund's rules for the electronic configuration of carbon.

3-147 (a), (b) and (d) are incorrect. Only (c) is correct.

3-149 (d) Fe^{3+} has five unpaired electrons.

3-151 5 unpaired electrons can be placed in a d subshell.

3-153 Atoms become smaller as we go across a row of the periodic table from left to right. Electrons that are added to the elements as we go from left to right across a row are added to the same shell, not to larger shells. Because the number of protons in the nucleus also increases, the core charge increases and therefore, the nucleus tends to have a stronger attraction to the electrons and the atoms become smaller.

3-155 Atomic radii decrease from left to right across a row and increase from top to bottom down a column. Aluminum will have the smallest radius.

3-157 The covalent radius increases down a column and decreases from left to right across a row in the periodic table. Phosphorus has the largest covalent radius.
The order is F < O < N < S < P.

3-159 As you go from Xe, adding one more electron to get to Cs, the electron is added to the next shell. By jumping to the next shell, the size of the atom will increase.

3-161 The radius of Pb^{2+} ion is larger than the radius of Pb^{4+} ion because the nucleus has a stronger attraction for the fewer electrons of the Pb^{4+} ion.

3-163 Ionic radii are used to estimate the distance between ions in compounds. These distances are important because properties of compounds are often dictated by how far one ion is from another.

3-165 The size of negative ions increases down a column. The order of increasing ionic radius is predicted to be: $H^- < F^- < Cl^- < Br^- < I^-$
The ionic radii (in nm) as given in the Appendix are:

$$\begin{array}{ccccccccc} 0.136 & < & 0.208 & < & 0.181 & < & 0.196 & < & 0.216 \\ F^- & < & H^- & < & Cl^- & < & Br^- & < & I^- \end{array}$$

The exception is hydrogen where the ratio of electrons to protons is much higher than those of the other elements in this series.

3-167 (e) Se^{2-}

3-169 (c) Be^{2+}

3-171 (b) P^{3-}. It has the smallest number of protons holding the electrons.

3-173 (a) Rb^{+}. It has the smallest positive charge in the same shell.

3-175 More highly charged cations have higher ionization energies. P^{4+} of the species listed will have the largest ionization energy.

3-177 The great increase in IE between the third and fourth ionization energies suggests an element in period three with 3 electrons in that period. This would be aluminum with electron configuration [Ne] $3s^2$ $3p^1$, response d.

3-179 (a) Of the elements presented, Na has the largest second ionization energy.

3-181 The order of increasing second ionization energy is
Mg < Be < Ne < Na < Li

3-183 Fe^{2+}: [Ar]$3d^6$ and Fe^{3+}: [Ar]$3d^5$. The $4s^2$ electrons are the first to be removed forming Fe^{2+}. In Fe^{3+}, an additional electron is removed from the 3d subshell to yield a stable ion with one unpaired electron in each orbital.

3-185 B: $\text{AVEE} = \dfrac{2(1.36) + 1(0.80)}{3} = 1.17$

F: $\text{AVEE} = \dfrac{2(3.88) + 5(1.68)}{7} = 2.31$

These are the same values found in Figure 3.31.

3-187 The AVEE is a weighted average of the ionization energies of all the valence electrons. Thus it is a measure of how easy it is to remove an electron from an atom. AVEE also shows how readily an atom will attract an electron. Atoms with a high value of AVEE have a greater ability to attract electrons.

3-189 AVEE decreases going down a particular column on the period table. This is because the ionization energy decreases going down the chart.

3-191 At the top of column VA, N has an AVEE of 1.82 MJ/mol, which is quite high. AVEE also shows how readily an atom will attract an electron to As, the AVEE has dropped to 1.26 which is in the range of semimetals. Sb also has an AVEE indicative of a semimetal. Below Sb, the AVEE of Bi would be an even lower AVEE, indicating that it would be a metal.

3-193 AVEE increases from left to right. As AVEE increases the metallicity decreases.

3-195 Bi < Pb < Au < Ba

3-197 (a) P (b) As (c) Sn (d) Ga

3-199 Volume of a sphere = $V = \frac{4}{3}\pi r^3$

Compare the $(\text{radii})^3$.

$$\frac{r^3_{nucleus}}{r^3_{atom}} = \frac{\left(10^{-14}\right)^3}{\left(10^{-10}\right)^3} = 10^{-12}$$

3-201 Increasing Ionization Energy: $O^{2-} < F^- < Ne < Na^+ < Mg^{2+}$
Increasing Radius: $Mg^{2+} < Na^+ < Ne < F^- < O^{2-}$
Increasing Ionization Energy: $Na < Mg < O < F < Ne$
Increasing Radius: $Ne < F < O < Mg < Na$

3-203 The shell model and quantum mechanical model of oxygen predict that electrons in shells farther from the nucleus require less energy to be removed. This is consistent with the first ionization energy data found in Table 3.3 and the photoelectron data for the third peak of oxygen found in Table 3.4. The shell model predicts that all electrons in the n=2 shell should require the same amount of energy to be removed. The photoelectron data found in Table 3.4 is not consistent with this idea. This is where the concept of subshells is introduced.

3-205 Relative ionization energies for an isoelectronic series depend on the total number of protons present: the greater the number of protons, the more strongly the outermost electron is held, and the higher the first ionization energy. In this case, the order of increasing ionization energy is S^{2-}(16 protons) < Ar(18 protons) < K^+(19 protons). According to the data found in Table 3.8, in an isoelectronic series, the cations are smaller than the neutral atoms which are in turn smaller than the anions. Thus, the order of increasing radii would be $K^+ < Ar < S^{2-}$.

3-207 (a) C, 2 unpaired electrons (b) N, 3 unpaired electrons (c) O, 2 unpaired electrons (d) Ne, 0 unpaired electrons (e) F, 1 unpaired electron

3-209 In the Stern-Gerlach experiment beams of atoms will interact with a magnetic field and split into two separate beams if they possess unpaired electrons. The magnetic moment is related to the number of unpaired electrons. Since He and Ne do not have a magnetic moment, we would expect that beams of these atoms would not be split. Beams of the other atoms should be split in two because they have magnetic moments.

3-211 Element Z is diamagnetic and typically forms +2 cations. This is consistent with Z having 2 electrons in its outermost shell. Group 2A matches this description. Since Z has the next to lowest ionization energy in its group and ionization decreases down a group, Z must be the next to last element in Group 2A: Ba. The compounds formed would be BaO and $BaCl_2$.

3-213

Atom	Chemical Symbol	Magnetic field behavior	First IE	Atomic Radius	Number of PES Peaks	Core Charge	AVEE	Number of Valence Electrons
X	Ne	not deflected	2.08	0.070	3	+8	2.0	8
Y	Na	deflected	0.50	0.16	4	+1	0.50	1
Z	Mg	not deflected	0.79	0.14	4	+2	0.8	2

3-215 (a) Cl. The atom is in the group VIIA with a core charge of +7 and very high AVEE indicates it is high up in that column. And the covalent radius matches other atoms in the third row.

(b) Cl: $1s^22s^22p^63s^23p^5$

(c) Cl^-: $1s^22s^22p^63s^23p^6$. The radius of Cl^- would be larger since there is one more electron, but the same core charge.

(d) It is easier to ionize Cl^- than it is Cl, because the number of electrons is larger and the core charge is the same.

(e) AVEE (Cl) < AVEE (Ar). AVEE increases to the right of the periodic table.

(f) This element is F. With a higher AVEE, it means the electrons are more tightly bound, probably in a smaller shell, indicating a smaller covalent radius.

(g) 7

3-217 (a) (ii) matches the ionization energies; two low energies, followed by two higher energies from the next shell.

(b) 2 the two in the outer shell.

(c) +2

(d) You are taking an electron from something which is already a positive ion.

Chapter 4
The Covalent Bond

Note to students:

There are several schools of thought on what constitutes the best Lewis structures for molecules. The best evidence, of course, is experimental data. In the absence of data, some chemists prefer to minimize the formal charge in the structure. Thus, for example, for SO_2, the structure can be drawn with one or two double bonds. In one case, sulfur has 10 valence electrons and no formal charge; in the other, sulfur has eight valence electrons and a formal charge of +1. Without additional information, it is not possible to feel confident of the correct structure. We have generally chosen not to use expanded octets in writing the structures of this text. A recent study[1] of Lewis structures suggests that multiple bonds (expanded octets) may not be the best representation of the structures of many molecules commonly drawn this way.

4-1 Valence electrons are the electrons in an atom's outermost shell. This means they are the electrons on an atom that were not present in the previous Group VIIIA (Group18) element, ignoring filled d or f subshells.

4-3 (a) Fe has 8 valence electrons.
(b) Cu has 11 valence electrons.
(c) Bi has 5 valence electrons.
(d) I has 7 valence electrons.

4-5 Na^+, Mg^{2+}, Al^{3+}, and Sc^{3+} all have 8 valence electrons. All of these elements lose electrons to achieve an octet of electrons.

4-7 The **octet rule** refers to Lewis's discovery that main-group elements will gain or lose electrons until they have eight electrons in their outermost shell.

4-9 For main-group elements and alkaline metals the group number is the same as the number of valence electrons.

4-11 In the covalent bond of F_2 there is a single pair of electrons shared between the two fluorine atoms. In the O_2 molecule there are two pairs of electrons shared between the two atoms. However, in both molecules there are eight valence electrons which surround each atom.

4-13 When two atoms are brought together the electrons on atom A would be attracted to the nucleus of atom B and *vice versa*. There would also exist an equal and opposite repulsion between the two nuclei and the separate electrons. These two forces would appear to negate each other as shown in Figure 4.4. However, if the shared electrons between the two atoms were specifically placed in the region between the two atoms the repulsions due to the nuclei can be minimized.

4-15 A **bonding domain** is a region of space which contains two spin paired electrons which are being shared between two atoms. The domain extends over both atoms, but effectively concentrates the electron density in the space between the atoms.

[1] L. Suidan, J.K. Badenhoop, E.D. Glendenins, and F. Weinhold, J. Chem. Ed. 72 583(1995).

4-17 According to the model for the electron configurations of an atom, two and only two electrons can share a common region of space. The two electron occupants have mutual access to the attraction of the two bound nuclei (and find themselves in an environment much more favorable than that existing in the isolated atoms). Thus, the unpaired valence electrons of two chlorine atoms come together to form the covalent bond of Cl_2.

4-19

3 H· + ·N̈· ⟶ H:N̈:H
H

Nonbonding Domain

Bonding Domains

4-21 Each atom should have eight electrons surrounding it to satisfy the octet rule.

4-23 (c)

4-25 In general the element with the lowest AVEE will be the central atom. Since S is directly below O on the periodic table we automatically know that it has a lower AVEE than oxygen therefore structure (a) is the best.

4-27

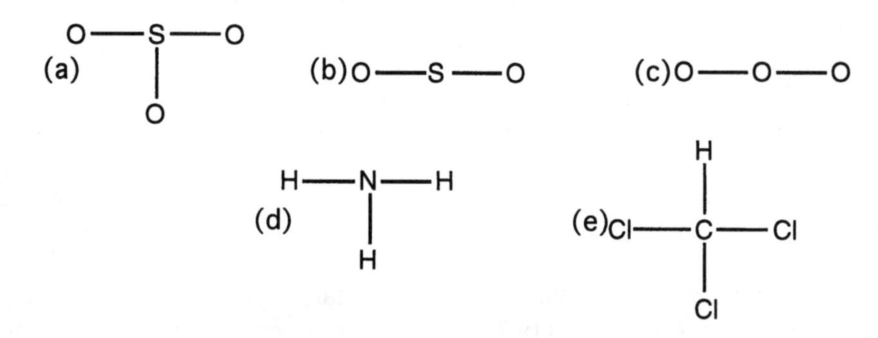

4-29 (a) 8 for Kr + 2(7) for F = 22 valence electrons
(b) 6 for S + 4 (7) for F = 34 valence electrons
(c) 4 for Si + 6(7) for F + 2 for charge = 48 valence electrons
(d) 4 for Zr + 7(7) for F + 3 for charge = 56 valence electrons

4-31

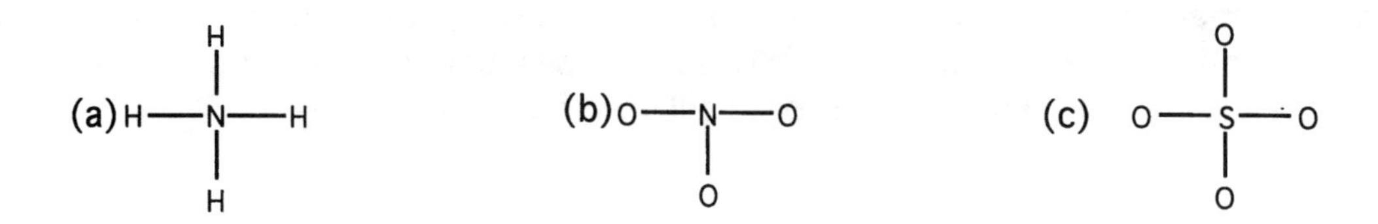

4-33

(a) NO_3^- (b) SO_3^{2-} or

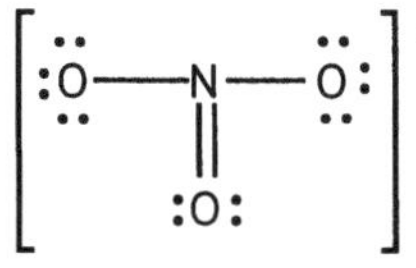

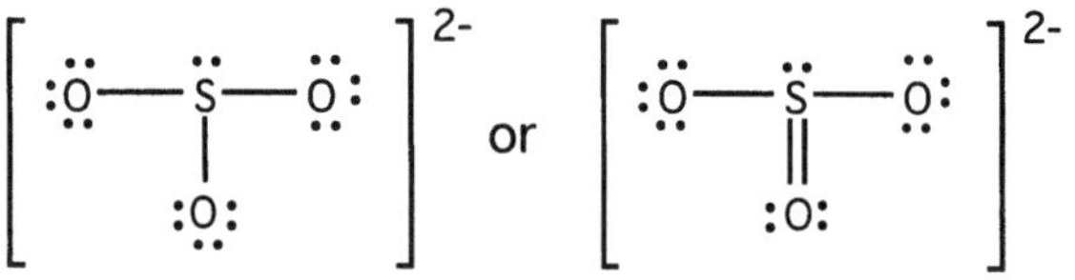

The structure with the minimum number of formal charges is generally preferred by some chemists.

(c) CO_3^{2-} (d) NO_2^+

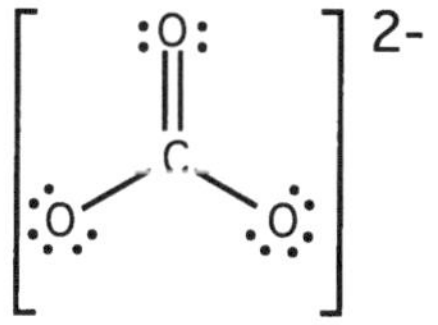

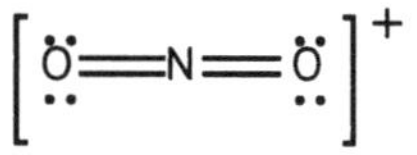

4-35

(a) N_2O

I II

Structure I is preferred.

(b) N_2O_3

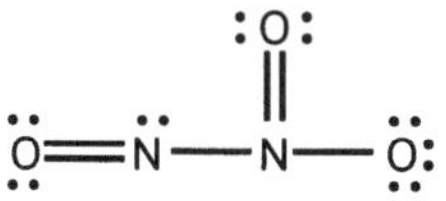

4-37 In the molecule N_2O_5, there are 40 valence electrons available. A structure that contains $O_2N\text{-}NO_3$ requires 42 valence electrons. The correct Lewis structure is:

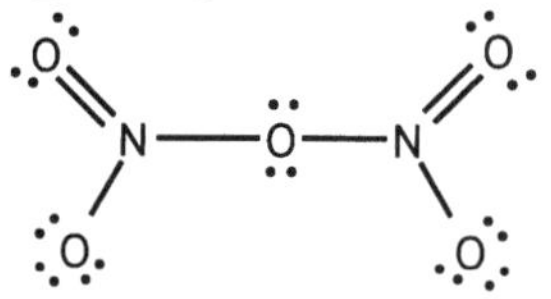

4-39

(a) SO_2 (b) SO_3 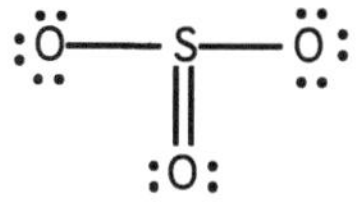 (c) SO_3^{2-} 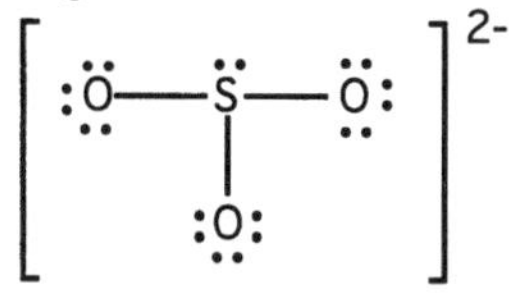 (d) SO_4^{2-}

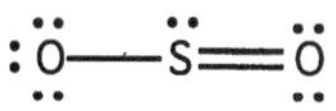

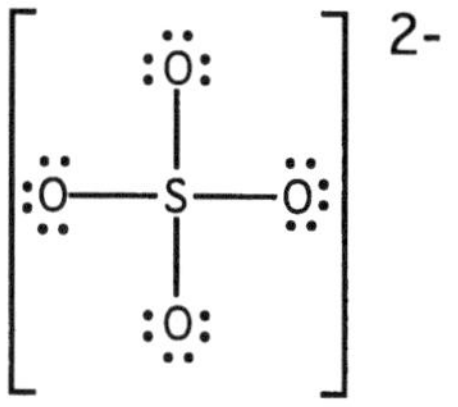

Structures (a),(b), and (d) have been shown to be of more significance than the expanded octet structures. See reference 1 at the beginning of this problem set.

4-41 The N_2 molecule has 10 valence electrons.

(a) CO has 10 valence electrons.
(b) NO has 11 valence electrons.
(c) CN^- has 10 valence electrons.
(d) NO^+ has 10 valence electrons.
(e) NO^- has 12 valence electrons.

CO, CN^-, and NO^+ have the same electronic configuration as the N_2 molecule.

4-43 (a) BF_3

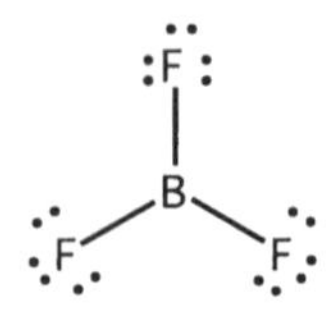

Boron is surrounded by six valence electrons. This is an exception to the octet rule.

(b) H_2CO

Formaldehyde is not an exception to the octet rule.

(c) XeF_4

Xenon is surrounded by twelve valence electrons. This is an exception to the octet rule.

(d) IF_3

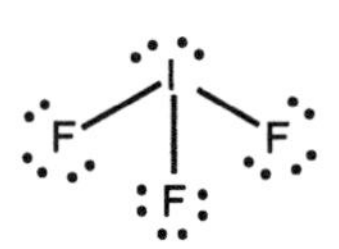

Iodine is surrounded by 10 valence electrons. This is an exception to the octet rule.

4-45

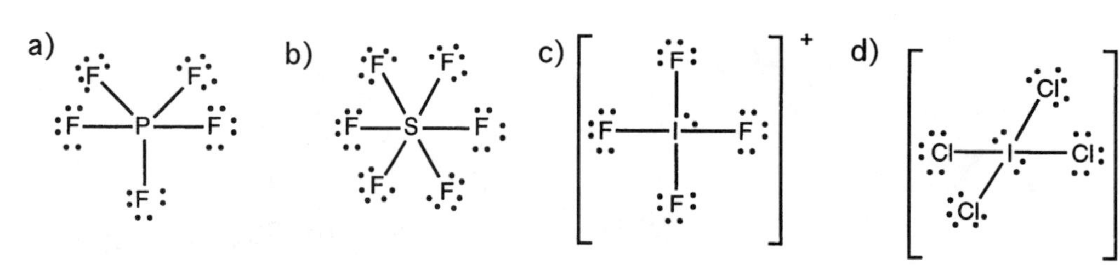

None of these structures obey the octet rule.

4-47 The shorter the bond length the higher the bond order; therefore N_2 is a triple bond, HNNH is a double bond and H_2NNH_2 is a single bond.

4-49 Since the bond lengths are about the same, the bond orders and hence the bond strengths must also be similar.

4-51 The resonance structures of SCN^- are:

$$[S=C=N]^- \longleftrightarrow [S\equiv C-N]^- \longleftrightarrow [S-C\equiv N]^-$$

4-53 Responses (b) and (c) have no resonance structures.

4-55 Since AVEE increases from left to right across in the periodic table, electronegativity also increases. Moving from left to right along a row in the periodic table metallic character decreases and covalent character increases. Since covalent radii are smaller than metallic radii, it is harder to remove an electron from an atom that has a greater covalent character i.e., the electron is closer to the nucleus and more tightly held. This is also consistent with the AVEE data rationalizing the trend of increasing electronegativity.

4-57 Response (b) is correct.

4-59 (a)

:F̤̈—F̤̈:

$$\delta_F = G_F - N_F - B_F\left(\frac{EN_F}{EN_F + EN_F}\right)$$

$$\delta_F = 7 - 6 - 2\left(\frac{4.19}{4.19 + 4.19}\right) = 7 - 6 - 1 = 0$$

(b)

H—F̤̈:

$$\delta_F = 7 - 6 - 2\left(\frac{4.19}{4.19 + 2.30}\right) = 7 - 6 - 1.29 = -0.29$$

(c)

:C̤̈l—F̤̈:

$$\delta_F = 7 - 6 - 2\left(\frac{4.19}{4.19 + 2.87}\right) = 7 - 6 - 1.19 = -0.19$$

4-61

:B—F̤̈:

$$\delta_F = 7 - 6 - 2\left(\frac{4.19}{4.19 + 2.05}\right) = -0.34$$

$$\delta_B = 7 - 6 - 2\left(\frac{2.05}{4.19 + 2.05}\right) = +0.34$$

4-63

$$\delta_O = 6 - 4 - 4\left(\frac{3.61}{3.61 + 2.54}\right) = -0.35$$

$$\delta_H = 1 - 0 - 2\left(\frac{2.30}{2.30 + 2.54}\right) = +0.05$$

4-65 (a) H, lower electronegativity than C.
(b) C, lower electronegativity than Cl.
(c) B, lower electronegativity than H.
(d) N, lower electronegativity than O.
(e) N, lower electronegativity than both Cl and O and it's between the two.

4-67 The formal charge on an atom is the difference between its number of valence electrons and the number of electrons the atom formally has in the Lewis structure. This is calculated by dividing the electrons in each covalent bond between the atoms (one to each atom) and then comparing the number of electrons that are now formally assigned to each atom with the number of valence electrons on the neutral atom.

(a) HBr

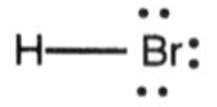

H—Br:

Formal charge (Br) = 7 valence e^- in neutral atom - 7 e^- formally assigned = 0

(b) Br_2

:Br—Br:

Formal charge (Br) = 7 valence e^- in neutral atom - 7 e^- formally assigned = 0

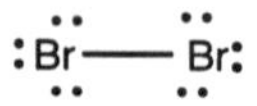

(c) HOBr

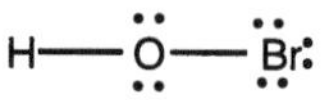

H—O—Br:

Formal charge (Br) = 7 valence e^- in neutral atom -7 e^- formally assigned = 0

(d) BrF_5

Formal charge (Br) = 7 valence e^- in neutral atom -7 e^- formally assigned = 0

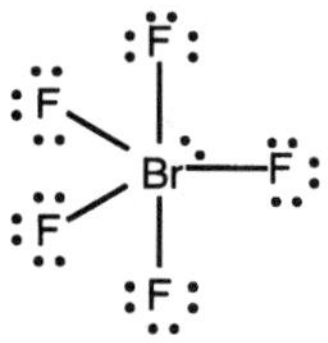

4-69 The formal charge on an atom is the difference between its number of valence electrons and the number of electrons the atom formally has in the Lewis structure. This is calculated by dividing the electrons in each covalent bond between the atoms (one to each atom) and then comparing the number of electrons that are now formally assigned to each atom with the number of valence electrons on the neutral atom.

(a) N_2O

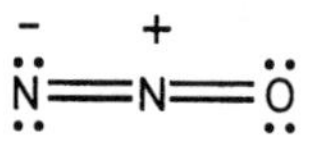

$\overset{-}{N}=\overset{+}{N}=O$

For the most important resonance structure

Formal charge on the end (N) = 5 valence e^- in neutral atom -5 e^- formally assigned = 0

Formal charge on middle (N) = 5 valence e^- in neutral atom -4 e^- formally assigned = +1

(b) N_2O_3 Formal charge (N_1) = 5 valence e^- in neutral atom - 5 e^- formally assigned =0

Formal charge (N_2) = 5 valence e^- in neutral atom -4 e^- formally assigned = +1

O=N_1—N_2—O: (with :O: double-bonded to N_2)

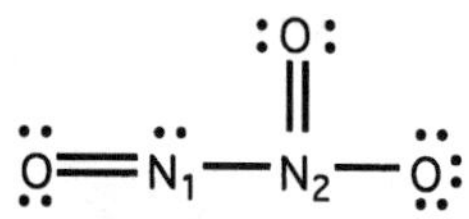

(c) N_2O_5

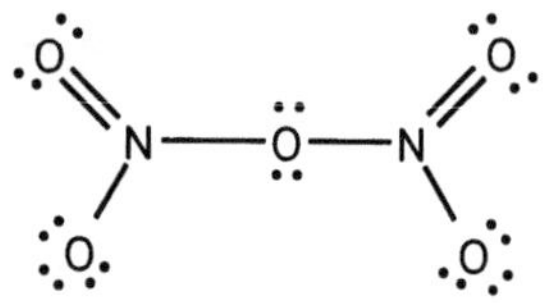

Formal charge (N) = 5 valence e^- in neutral atom -4 e^- formally assigned = +1

4-71 The formal charge on an atom is the difference between its number of valence electrons and the number of electrons the atom formally has in the Lewis structure. This is calculated by dividing the electrons in each covalent bond between the atoms (one to each atom) and then comparing the number of electrons that are now formally assigned to each atom with the number of valence electrons on the neutral atom.

$S_2O_3^{2-}$

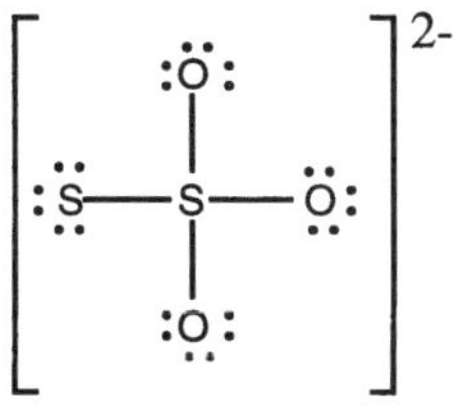

Formal charge end (S) = 6 valence e^- in neutral atom -7 e^- formally assigned = -1
Formal charge middle (S) = 6 valence e^- in neutral atom -4 e^- formally assigned = +2
Formal charge each (O) = 6 valence e^- in neutral atom -7 e^- formally assigned = -1
or

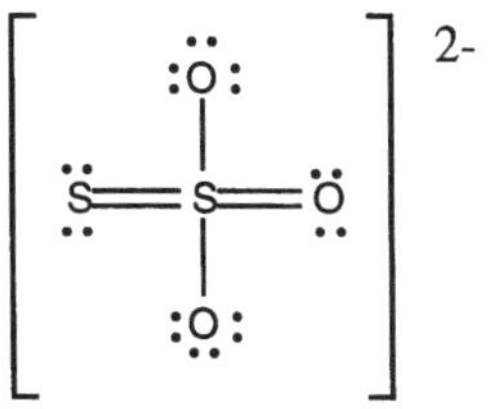

Formal charge end (S) = 6 valence e^- in neutral atom - 6e^- formally assigned = 0
Formal charge middle (S) = 6 valence e^- in neutral atom - 6 e^- formally assigned = 0
Formal charge (single bond O) = 6 valence e^- in neutral atom -7 e^- formally assigned = -1
Formal charge (double bond O) = 6 valence e^- in neutral atom - 6 e^- formally assigned = 0

4-73

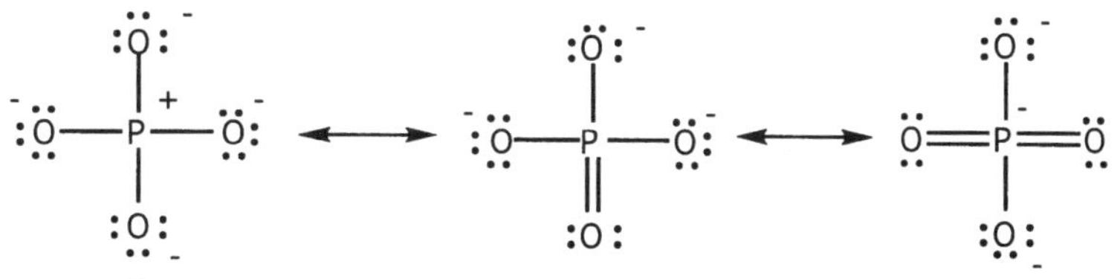

I II III

Structure II would be the preferred structure. The formal charge on each atom is minimized, and the negative formal charge is on the more electronegative oxygen. Experimental measurements of bond lengths could be used to confirm this. See however reference 1.

4-75

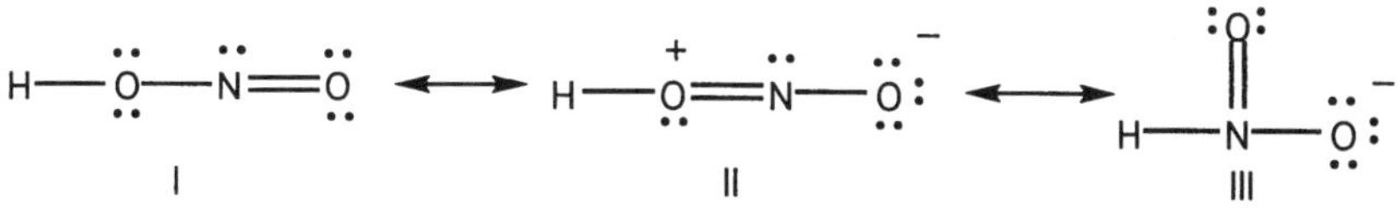

I II III

The first structure, HONO, has two possible Lewis structures. The one on the left, I, has no formal charge on any atom. There is only one structure for HNO_2. This structure has a formal charge on N and one O. Structure I is best.

4-77 The three structures are

a) $O{=}C{=}N^-$ b) $^+O{\equiv}C{-}N^{2-}$ c) $^-O{-}C{\equiv}N$

Oxygen is the most electronegative atom in this molecule so a negative charge should be placed at that atom, making structure c) the best structure.

4-79 a) 3 domains, trigonal planar, 120°.

:O:
||
C
:O: :O:

b) 4 domains, tetrahedral, 109.5°

[:O: O—S—O :O:] 2-

c) 6 domains, octahedral, 90°

[F F F—P—F F F] -

d) 5 domains, trigonal bipyramidal, 90° and 120°

F F
F—As—F
F

4-81 (a) $\underline{N}H_4^+$

four bonding domains,
bond angles of 109.5°

(b) $H\underline{C}Cl_3$

four bonding domains,
bond angles of 109.5°

(c) $\underline{Be}H_2$

two bonding domains,
bond angles of 180°

H—Be—H

(d) $O\underline{C}Cl_2$

three bonding domains,
bond angles of 120°

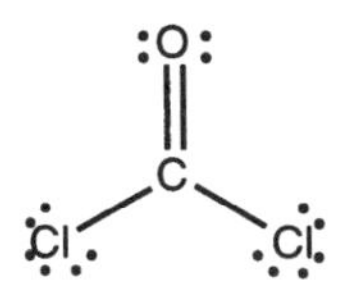

(e) $O\underline{C}S$

two bonding domains,
bond angles of 180°

4-83 (a) OH^-

Three pairs of nonbonding electrons

(b) O_2

Two pairs of nonbonding electrons

(c) CO_3^{-2}

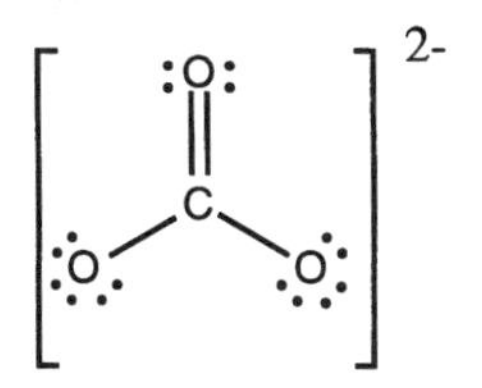

Eight pairs of nonbonding electrons

(d) Br^-

Four pairs of nonbonding electrons

(e) NH_3

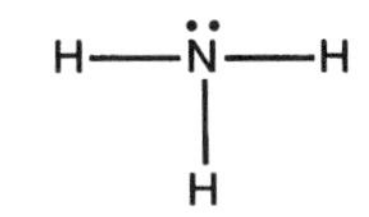

One pair of nonbonding electrons

All species have at least one pair of nonbonding electrons

4-85

Bonding domains	Nonbonding domains	Arrangement	Molecular geometry
2	0	linear	linear
4	0	tetrahedral	tetrahedral
2	2	tetrahedral	bent
5	0	trigonal bipyramidal	trigonal bipyramidal
3	2	trigonal bipyramidal	T-Shaped
5	1	octahedral	square pyramidal
4	2	octahedral	square planar

4-87 (a) PO_4^{3-}

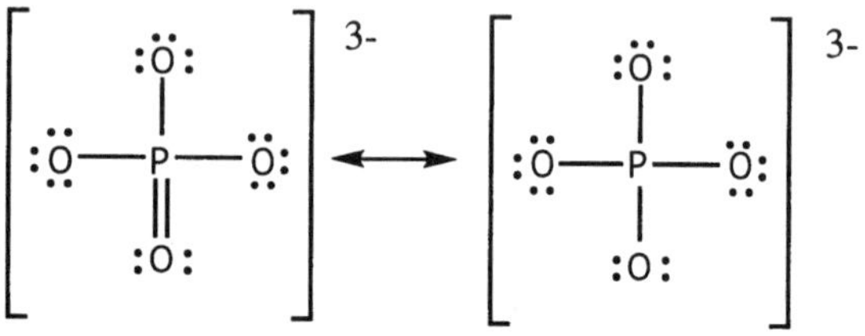

Number of bonding domains= 4
Number of nonbonding domains = 0
Geometry = tetrahedral

(b) SO_4^{2-}

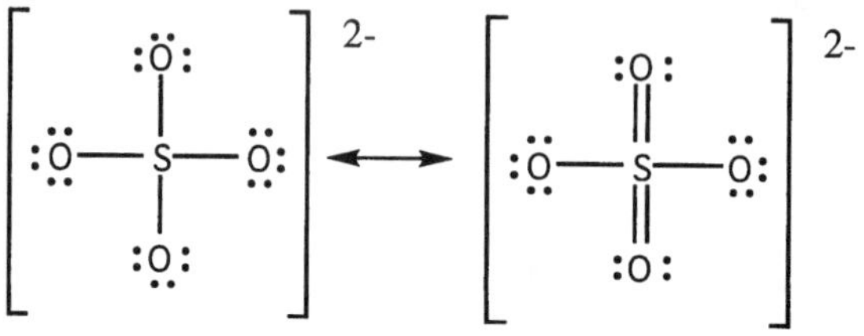

Number of bonding domains= 4
Number of nonbonding domains = 0
Geometry = tetrahedral

(c) XeO_4

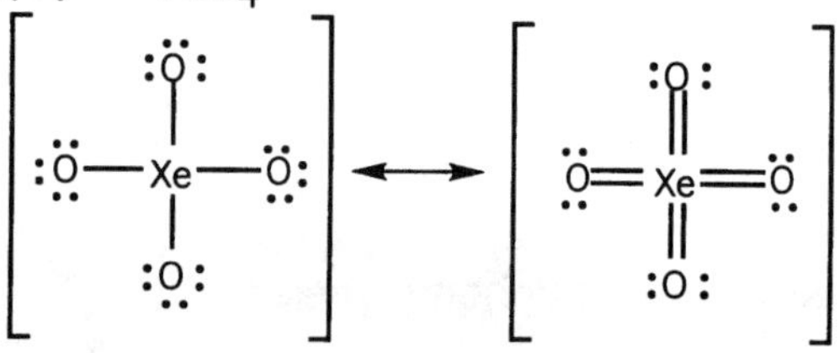

Number of bonding domains = 4
Number of nonbonding domains = 0
Geometry = tetrahedral

(d) MnO_4^-

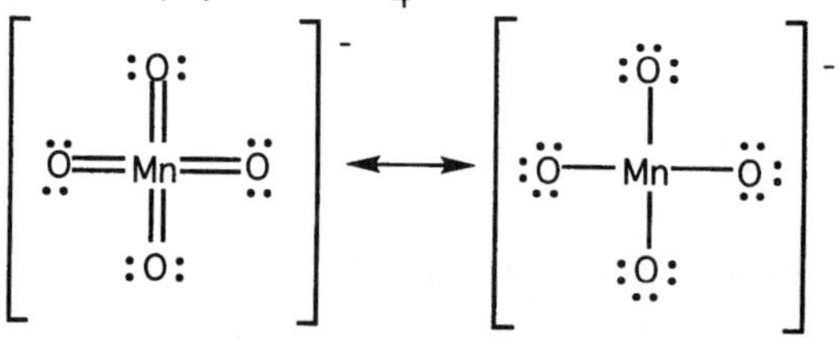

Number of bonding domains = 4
Number of nonbonding domains = 0
Geometry = tetrahedral

4-89 (a) SF_3^+

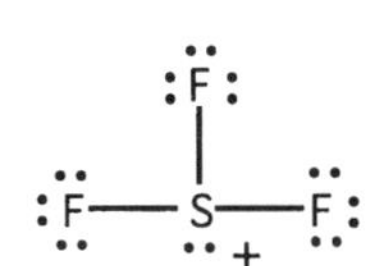

Number of bonding domains= 3
Number of nonbonding domains = 1
Geometry = trigonal pyramidal

(b) SF_4

Number of bonding domains = 4
Number of nonbonding domains = 1
Geometry = see-saw

(c) SF_5^-

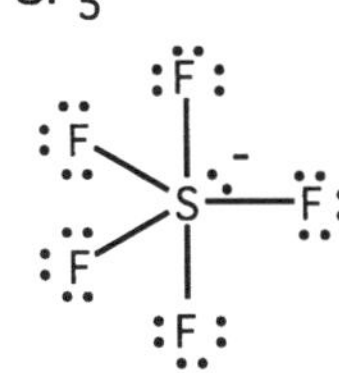

Number of bonding domains= 5
Number of nonbonding domains = 1
Geometry = square pyramidal

(d) SF_6

Number of bonding domains= 6
Number of nonbonding domains = 0
Geometry = octahedral

4-91 (a) N_2O

Number of bonding domains = 2
Number of nonbonding domains= 0
Geometry = linear

(b) NO_2^-

Number of bonding domains = 2
Number of nonbonding domains= 1
Geometry = bent

(c) NO_3^-

Number of bonding domains = 3
Number of nonbonding domains = 0
Geometry = trigonal planar

4-93 Response (a). XeF_3^+ has a T-shaped molecular geometry.

4-95

SF_4	seesaw	N_2O	linear
CH_4	tetrahedral	NO_2	bent
CO_2	linear	PCl_4^+	tetrahedral
H_2O	bent	PCl_4^-	seesaw
BeH_2	linear		

Response (c) CO_2 and BeH_2 are linear.

4-97 (a) SO_3 trigonal planar
(b) SO_3^{2-} trigonal pyramidal
(c) NO_3^- trigonal planar
(d) PF_3 trigonal pyramidal
(e) BH_3 trigonal planar

Responses (a),(c),and (e) have planar geometry.

4-99 (a) C_2H_2 linear
(b) CO_2 linear
(c) NO_2^- bent
(d) NO_2^+ linear
(e) H_2O bent

Responses (a), (b), and (d) have linear geometry.

4-101

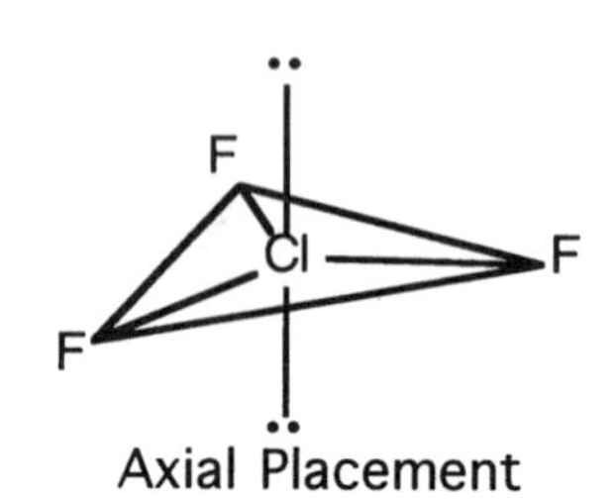

Axial Placement

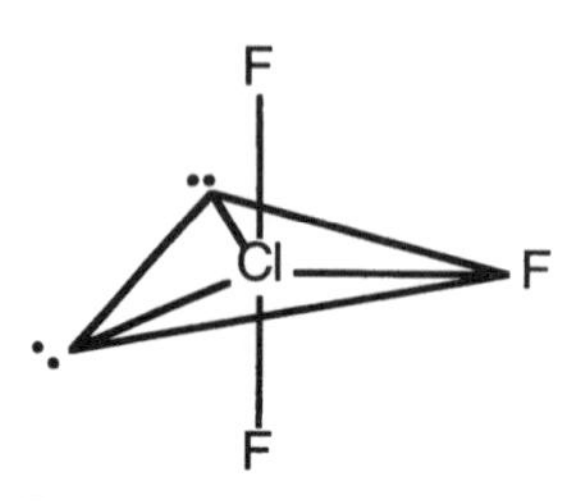

Equatorial Placement

With the nonbonding domains in the equatorial position, there are four unfavorable 90° bonding-nonbonding interactions. With the nonbonding domains in the axial positions, there are 6 unfavorable 90° bonding-nonbonding interactions. The placement of the nonbonding domains in the equatorial position gives a more stable structure.

4-103

Bonding Domains	Nonbonding Domains	Bond Angle
2	0	180°
5	0	90° and 120°
3	0	120°
2	1	<120°
2	3	180°
4	0	109.5°
3	1	<109.5°
2	2	<109.5°
2	3	180°
3	2	<90°
4	2	90°

4-105 a) both C 109.5°, O <109.5°
b) 109.5°
c) O <109.5°, N <120°
d) H_3C < 109.5°, O=C 120°
e) 120°

4-107 a) b)

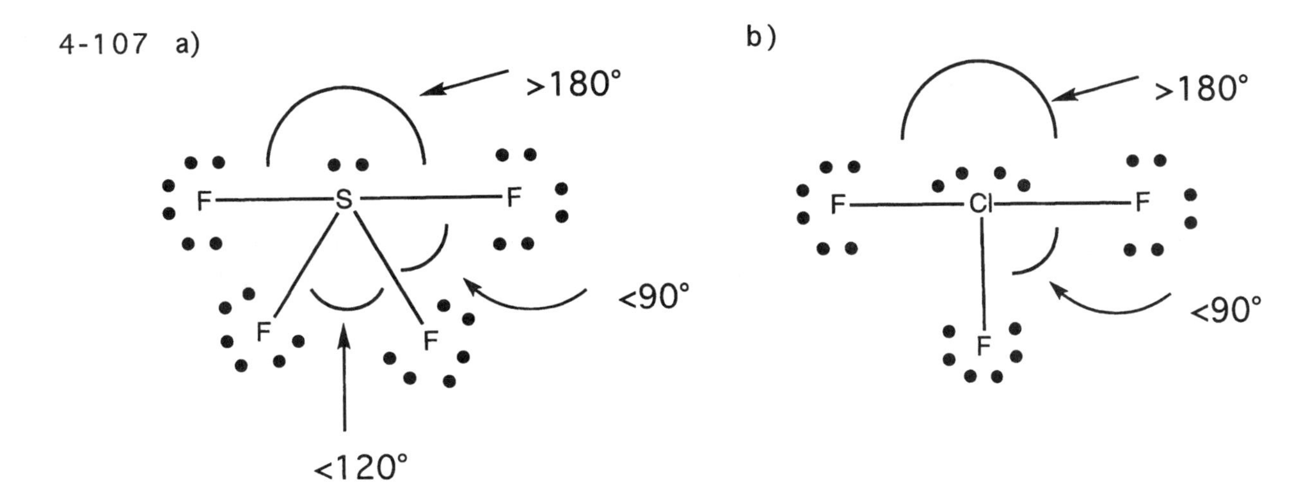

c) d)

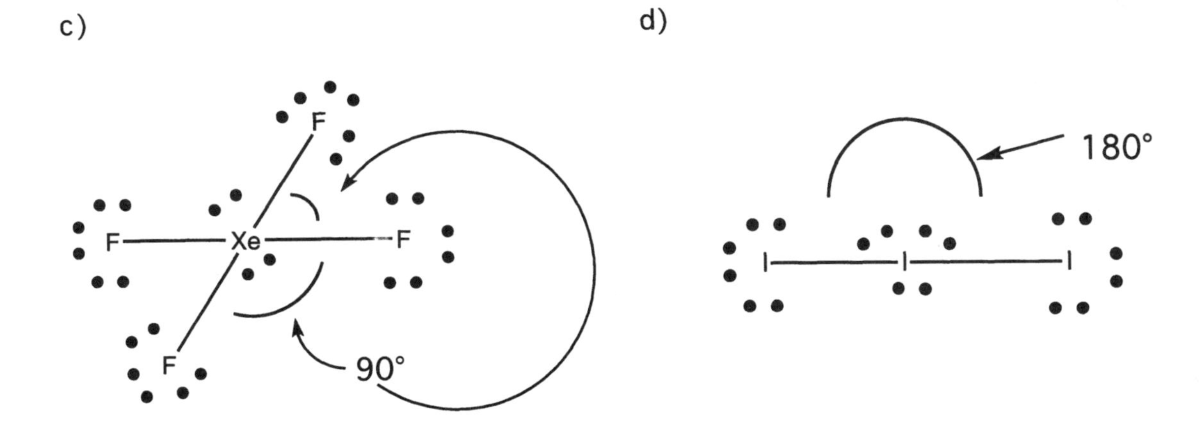

4-109 The carbon-sulfur bond is a polar bond. In CS_S the dipoles cancel because the molecule is symmetrical. Therefore CS_2 is not a polar molecule. The dipoles created by the bonds in OCS, however, do not cancel.

4-111 All responses are polar. Response (c) is the least polar. Although dimethyl ether,(c), has a structure similar to that of water, the C-O bond is not as polar as the H-O bond.

4-113 The Lewis structures of thionyl chloride ($SOCl_2$) and sulfuryl chloride (SO_2Cl_2) are:

:Cl—S—Cl: (S=O:) :Cl—S—Cl: (O=S=O) ⟷ :Cl—S—Cl: (O—S—O)

thionyl chloride sulfuryl chloride

Thionyl chloride has a dipole moment along the S-O bond. Therefore, it is a polar molecule. There are two possible structures that could be drawn for sulfuryl chloride, and both structures have a tetrahedral arrangement. Since there are two types of polar bonds in the molecule, S-Cl and S-O, the individual bond dipoles do not cancel. Sulfuryl chloride is a polar molecule.

4-115

Phosphorous with a formal charge of +1 is capable of interactions with an oxygen, formal charge -1, of another PO_3^- unit to polymerize. Similarly SO_3 would be expected to polymerize.

4-117

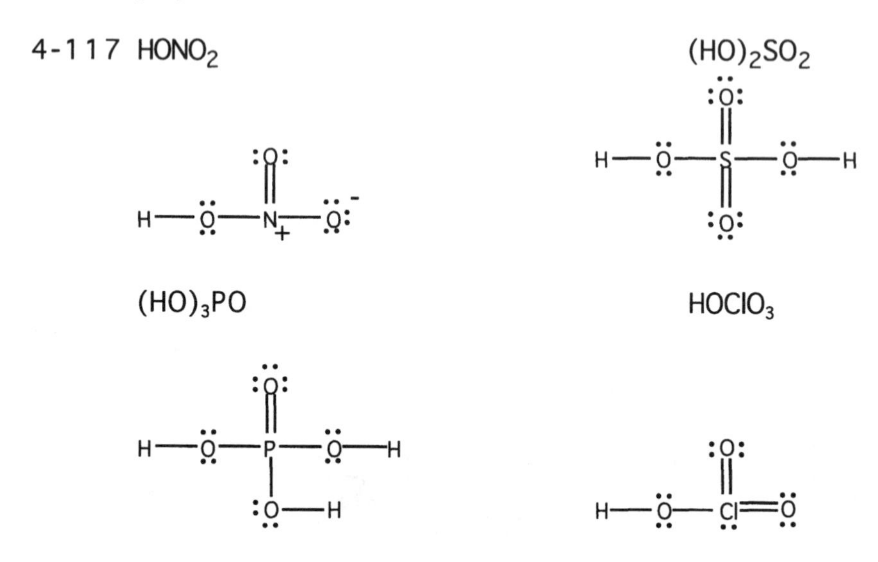

4-119 In the molecule with the Lewis structure,

:O—X=O

the central atom must have six valence electrons due to formal charge considerations. The element is S. Response (d).

4-121 For the molecule XF_6^{2-}, F will not double bond to the central atom. Therefore all bonds to the central atom will be single bonds. This means that the central atom must have 6 valence electrons for bonding to the fluorine atoms. The charge on the molecule is 2- so we have to subtract 2. The number of valence electrons on the central atom is 6-2 = 4. This narrows the choice to C or Si. Carbon cannot expand its valence shell to accept 6 bonds therefore it must be Si. The molecule is SiF_6^{2-}.

4-123 An octahedral molecular geometry in a molecule with four ligands such as XCl_4^-, requires the central atom to have 2 pairs of nonbonding electrons. The central atom must have 4 valence electrons for bonding to the four ligands. In addition the central atom needs another 4 valence electrons to account for the 2 pairs of nonbonding electrons. The molecule has a -1 charge, so we must subtract 1. The number of valence electrons on the central atom is 4 + 4 - 1 = 7. The central atom belongs to group VIIA.

4-125 Response (d). O_2^{2-} contains only a single bond.

4-127 A linear molecular geometry with two fluorine ligands and a -1 charge, XF_2^-, requires that the central atom have 2 valence electrons for bonding to the fluorine atoms. Since the molecule has a -1 charge, we must subtract 1. The number of valence electrons on the atom bonded to the fluorine is 2-1 = 1. The atom belongs to group IA. A second answer to this question can be found by assuming a trigonal bipyramidal arrangement of electron domains with 3 nonbonding domains at the equatorial positions and the fluorines attached to the axial positions. This would require 8e⁻ (6 nonbonding and 2 bonding) from the central atom. One electron is supplied by the charge leaving 7 electrons to be supplied by the central atom. The atom belongs to VIIA.

4-129

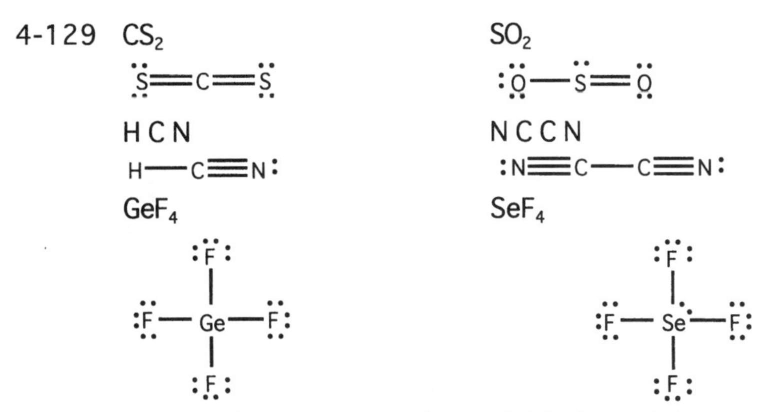

For a molecule to be polar, individual bond dipoles do not cancel each other. This is the case in SO_2, HCN, and SeF_4.

4-131 Bond length data for the SO_2 molecule would not provide enough information to decide which structure is correct. If the lengths of typical S-O and S=0 bonds were also available for comparison, then a choice could be made.

4-133(a) (b)

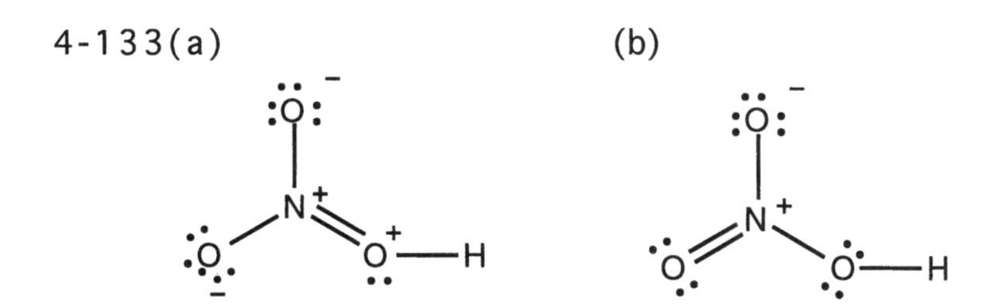

Structure (b) will contribute because it has less formal charge distribution than structure (a) which has more formal charges, but also a positive charge on an oxygen.

4A-1 Orbital overlap occurs when two orbitals, one from each atom engaged in a bond occupy the same place in space.

4A-3

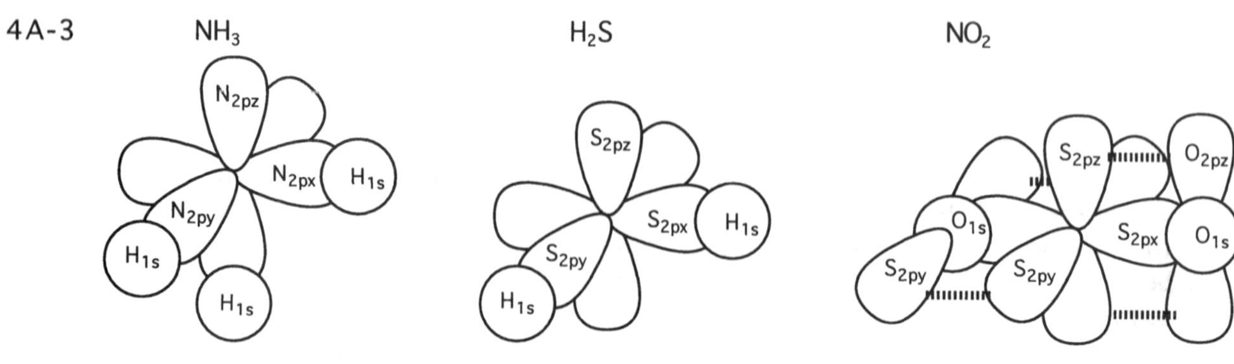

4A-5 (a) CH_4 sp^3
(b) H_2CO sp^2
(c) HCO_2^- sp^2

4A-7 The Lewis structure for the molecule ethylene is:

```
H       H
 \     /
  C===C
 /     \
H       H
```

The hybridization on each carbon is sp^2. Each C-H bond is σ (sp^2-s). One of the C-C bonds is a σ (sp^2-sp^2) and the other is a π (2p-2p). The molecule is planar.

The Lewis structure for the molecule acetylene is:

H—C≡C—H

The hybridization on each carbon is sp. The C-H bonds are σ (sp-s) bonds. One C-C bond is σ (sp-sp) and the other two C-C bonds are π (2p-2p) bonds. The molecule is linear.

4A-9 A σ molecular orbital is one that results from the head-on overlap of atomic orbitals, that is the electron density lies along the internuclear axis. A π molecular orbital is one that results from the sideways or parallel overlap of atomic orbitals. The electron density in a π molecular orbital does not lie on the internuclear axis.

4A-11 The stronger the interaction between a pair of atomic orbitals, the larger the difference between the energies of the bonding and antibonding orbitals formed. The σ_p molecular orbitals result from the head-on overlap of $2p_z$ atomic orbitals. This is a stronger interaction than the sideways overlap of the $2p_x$ and $2p_y$ atomic orbitals which results in the π_x and π_y molecular orbitals. Therefore the difference in energy between the σ_p and the σ^*_p will be greater than the difference in energy between the π_x π_y and π^*_x π^*_y molecular orbitals.

4A-13 The bond order (BO) for a diatomic molecule is calculated by placing the valence electrons in the molecular orbitals and counting those that are in bonding MO's and those that are in antibonding MO's. The equation for bond order is:

$$BO = \frac{(\text{no. of electrons in bonding MO's} - \text{no. of electrons in antibonding MO's})}{2}$$

(a) H_2 2 e^- $(\sigma_{1s})^2$ BO = 2/2 = 1

(b) C_2 8 e^- $(\sigma_{2s})^2 (\sigma^*_{2s})^2 (\pi_x)^2 (\pi_y)^2$

$BO = \frac{(6-2)}{2} = 2$

(c) N_2 10 e^- $(\sigma_{2s})^2 (\sigma^*_{2s})^2 (\pi_x)^2 (\pi_y)^2 (\sigma_p)^2$

$BO = \frac{(8-2)}{2} = 3$

(d) O_2 12 e^- $(\sigma_{2s})^2 (\sigma^*_{2s})^2 (\sigma_p)^2 (\pi_x)^2 (\pi_y)^2 (\pi^*_x)^1 (\pi^*_y)^1$

$BO = \frac{(8-4)}{2} = 2$

(e) F_2 14 e^- $(\sigma_{2s})^2 (\sigma^*_{2s})^2 (\sigma_p)^2 (\pi_x)^2 (\pi_y)^2 (\pi^*_x)^2 (\pi^*_y)^2$

$BO = \frac{(8-6)}{2} = 1$

4A-15 Calculate the bond order for the molecules. The molecule with the highest bond order is more stable.

O_2 12 e^- $(\sigma_{2s})^2 (\sigma^*_{2s})^2 (\sigma_p)^2 (\pi_x)^2 (\pi_y)^2 (\pi^*_x)^1 (\pi^*_y)^1$

BO = 2

O_2^{2-} 14 e^- $(\sigma_{2s})^2 (\sigma^*_{2s})^2 (\sigma_p)^2 (\pi_x)^2 (\pi_y)^2 (\pi^*_x)^2 (\pi^*_y)^2$

BO = 1

O_2 is more stable than O_2^{2-}.

4A-17 A molecule is paramagnetic if it contains unpaired electrons. The peroxide ion, O_2^{2-} has the electron configuration:

$(\sigma_{2s})^2 (\sigma^*_{2s})^2 (\sigma_p)^2 (\pi_x)^2 (\pi_y)^2 (\pi^*_x)^2 (\pi^*_y)^2$

All the electrons are paired. The peroxide ion is not paramagnetic.

4A-19 A molecule is paramagnetic if it contains unpaired electrons. It is diamagnetic when all electrons are paired.

(a) HF 8 e^- $(\sigma_{s\text{-}p})^2$

Because the hydrogen and fluorine atoms are so different, the MO's for this molecule do not follow the general pattern. The $\sigma_{s\text{-}p}$ MO is occupied by the 1s electron of hydrogen and one of the 2p electrons of fluorine. The other valence electrons are localized on the fluorine atom and do not participate in the MO bonding scheme.
All electrons are paired. HF is diamagnetic.

(b) CO 10 e^- $(\sigma_{2s})^2 (\sigma^*_{2s})^2 (\sigma_p)^2 (\pi_x)^2 (\pi_y)^2$

All electrons are paired. CO is diamagnetic.

(c) CN^- 10 e^- $(\sigma_{2s})^2 (\sigma^*_{2s})^2 (\sigma_p)^2 (\pi_x)^2 (\pi_y)^2$

All electrons are paired. CN^- is diamagnetic.

(d) NO 11 e^- $(\sigma_{2s})^2 (\sigma^*_{2s})^2 (\sigma_p)^2 (\pi_x)^2 (\pi_y)^2 (\pi^*_x)^1$

NO has an unpaired electron. It is paramagnetic.

(e) NO^+ 10 e^- $(\sigma_{2s})^2 (\sigma^*_{2s})^2 (\sigma_p)^2 (\pi_x)^2 (\pi_y)^2$

All electrons are paired. NO^+ is diamagnetic.

Chapter 5
Ionic and Metallic Bonds

5-1 The elements of the third row of the periodic table in order of decreasing metallic character are: Na > Mg > Al > Si > P > S > Cl > Ar and in order of increasing nonmetallic character are: Na < Mg < Al < Si < P < S < Cl < Ar.
In general, elements decrease in metallic properties going from left to right across a row, and they increasingly take on non-metal characteristics moving from left to right across the same row. Along a diagonal band, elements such as B, Si, Ge, As, Sb, Te, and At are identified as semi-metals because they exhibit properties of both metallic and nonmetallic elements. In the third row of the periodic table the metals are Na, Mg, Al. Si is a semimetal. P, S, Cl, and Ar are nonmetals.

5-3 In order of increasing non-metallic character:
(a) Sr < Ge < Al < N
(b) Rb < K < Mg < Si
(c) Ge < As < P < N
(d) Al < B < N < F The correct sequence.

5-5 Sodium is said to be more metallic than lithium because of its ability to more readily lose electrons.

5-7 (a) K (b) Na (c) Ca (d) Ca

5-9 All the alkali metals have the same valence configuration of xs^1 and very small AVEE values. Thus it is easy for these atoms to form +1 ions.

5-11 Xenon

5-13 Since there is only one valence electron for all the elements in group I, the AVEE is the same as first ionization energy. As one goes down the column the number of shells increases. As the number of shells increases, the ionization energy of an electron from that shell decreases. So the ionization energy (and AVEE) will decrease from Li to Fr.

5-15 The second ionization energy for Li would be much higher than the first for two reasons. First, the removal of an electron from a +1 ion requires more energy than removing it from a neutral atom (see Coulomb's law). Second, the second ionization would remove an electron from a smaller shell than the first ionization. This means the second electron is removed from a smaller distance from the nucleus, which also increases the ionization energy.

5-17 Be^{+2}: [He]
Mg^{+2}: [Ne]
Ca^{+2}: [Ar]
Sr^{+2}: [Kr]
Ba^{+2}: [Xe]
Ra^{+2}: [Rn]

5-19 Na has a lower first ionization than Mg because they are filling electrons in the same shell, but Na has less protons in the nucleus.

5-21 As one goes down the column the number of shells increases. The more shells the larger the atomic size, therefore the atomic size increases from Be to Ra.

5-23 Aluminum

5-25 From B to Al the ionization energy and the AVEE decrease because the valence shell is larger for Al than for B, so it is easier to pull off these outer electrons.

5-27 Ga: [Ar]$4s^2 3d^{10} 4p^1$
Ga$^+$: [Ar]$4s^2 3d^{10}$
Ga^{+2}: [Ar]$4s^1 3d^{10}$
Ga^{+3}: [Ar]$3d^{10}$
The fourth ionization from Ga would be more difficult than the third, since the removal of an electron from a +2 ion requires more energy than removing it from a +3 ion (see Coulomb's law). Second, the fourth ionization would remove an electron from the filled *3d* subshell. This means the fourth electron is removed from a smaller distance from the nucleus, which also increases the ionization energy.

5-29 (a) As is in Group VA 5-8=-3 As^{3-}
(b) Te is in Group VIA 6-8=-2 P^{2-}
(c) Se is in Group VIA 6-8=-2 S^{2-}

5-31 The hydrogen of metal hydrides is a negative ion, H^- . The hydrogen atom in non-metal hydrides is assigned a +1 oxidation state, H^+. In the non-metal compounds of hydrogen, the hydrogen is covalently bound, which means that an electron pair is shared between the two bound atoms.

5-33 All the halogens have the same valence configuration of xp^5 and very large AVEE values. However, they also have high electron affinities, thus it is easy for these atoms to attach an electron to close their electronic shell and form –1 ions.

5-35 Ions of main-group nonmetal elements are typically negative ions because they typically have high values for AVEE and so do not for positive ions. They also have high electron affinities and may adopt the noble gas electronic configuration, therefore they typically form ions by adding electrons making them negative.

5-37 Hydrogen has an AVEE that is almost in the middle of all main-group elements. When it reacts with an element with a low AVEE, it behaves like a halogen (NaH, sodium hydride). When it reacts with a high AVEE it behaves like an alkali metal (HBr, hydrogen bromide).

5-39 Zn: [Ar]$4s^2 3d^{10}$. The most likely change on a zinc ion would be +2. Since the atom would lose the two *4s* electrons, and have a closed *3d* shell.

5-41 Ni: [Ar]$4s^2 3d^8$
Ni^{+2}: [Ar] $3d^8$

5-43 (a) Na_2O, NaH (b) MgO, MgH_2 (c) Al_2O_3, AlH_3 (d) SiO_2, SiH_4
(e) P_2O_5, PH_3
The element best matching the suggested compositions is magnesium.

5-45 Main-group elements matching the reactivities described are to be found in group IIIA.

5-47 The very high third ionization energy for magnesium (7733 kJ).

5-49 The most important factor is the electron configuration and the number of electrons that must be gained or lost to achieve a filled outer shell.

5-51 The product of the reaction of strontium metal with phosphorus should have the formula Sr_3P_2 .

5-53 (a) $2\ Na(s) + F_2(g) \rightarrow 2\ NaF(s)$
(b) $4\ Na(s) + O_2(g) \rightarrow 2\ Na_2O(s)$
(c) $2\ Na(s) + H_2(g) \rightarrow 2\ NaH(s)$
(d) $16\ Na(s) + S_8(s) \rightarrow 8\ Na_2S(s)$
(e) $12\ Na(s) + P_4(s) \rightarrow 4\ Na_3P(s)$

5-55 (a) ZnF_2 (b) AlF_3 (c) SnF_2 or SnF_4 (d) MgF_2 (e) BiF_3 or BiF_5

5-57 **Superoxides** are compounds which form when a very reactive alkali metal, such as potassium reacts with a 1:1 ratio, with the O_2 molecule to form KO_2.

5-59 The Lewis structures for

Rb Rb⁺ O^{2-}

˙Rb Rb :Ö:

For rubidium oxide and rubidium superoxide the structures are:

Rb :Ö: Rb :Ö—Ö: Rb

5-61 Ionic compounds are held together by electrostatic or Coulombic forces which hold together oppositely charged species.

5-63 The structures for Cl, Na and the compound NaCl are:

:C̤̈l˙ ˙Na :C̤̈l: Na

5-65 In an ionic compound the electrostatic forces between the ions hold the compound together. In a covalent compound it is the sharing of electrons to give each atom a full octet of electrons which holds the molecule together.

5-67 The unit cell represents the simplest repeating unit which when replicated in three dimensions will reproduce the structure of the crystal. Common properties of the unit cell are that opposite faces of all unit cells are parallel and the edge of a unit cell connects equivalent points.

5-69

5-71 Ionic and covalent bonds are similar in that the electrons of the bonded atoms are localized. That is, the electrons reside either on a particular ion (ionic bonding) or shared by two different atoms (covalent bonding).

5-73 A metallic bond is one in which a group of positive metal ions are held together by a mobile sea of negative electrons.

5-75 Atoms which engage in metallic bonding are typically large and have very low values of AVEE as a result. Furthermore they don't have enough valence electrons to possibly share enough electrons to fill their valence shell, therefore losing them to form a metallic bond takes very little energy.

5-77 In a metallic bond the electrons are not localized on a particular atom. In an ionic bond the electrons arė localized on a specific ion. In both cases it is the electrostatic forces which hold the species together.
In a covalent bond, while the electrons are not localized on an atom, they are localized between two atoms which share the electrons to fill their respective valence shells. However in the metallic bond, electrons are not localized at all and it is the electrostatic attraction between the metal cations and mobile electrons which create the metallic bond.

5-79 The electronegativity difference is the smallest for F and O. Response (d).

5-81 From the bond type triangle

(a) CO	covalent
(b) H_2O	polar covalent
(c) BeF_2	ionic/covalent
(d) $MgBr_2$	ionic
(e) AlI_3	ionic/covalent
(f) ZnS	ionic/covalent
(g) $CdLi$	metallic

5-83 none of these

5-85 (a) and (d)

5-87 Using the guide that ionic substances generally result when metals combine with non-metals.

(a)	OF_2	covalent
(b)	CS_2	covalent
(c)	MgO	ionic
(d)	ZnS	ionic

Using a bond type triangle as a guide.
(a) covalent
(b) covalent
(c) ionic
(d) covalent/ionic

5-89 A smaller triangle such as that of this problem does not allow a clean division into three areas of bonding. The possible binary compounds and their classifications are given below. The dividing lines can be found by plotting these compounds on the triangle along with the given classifications.

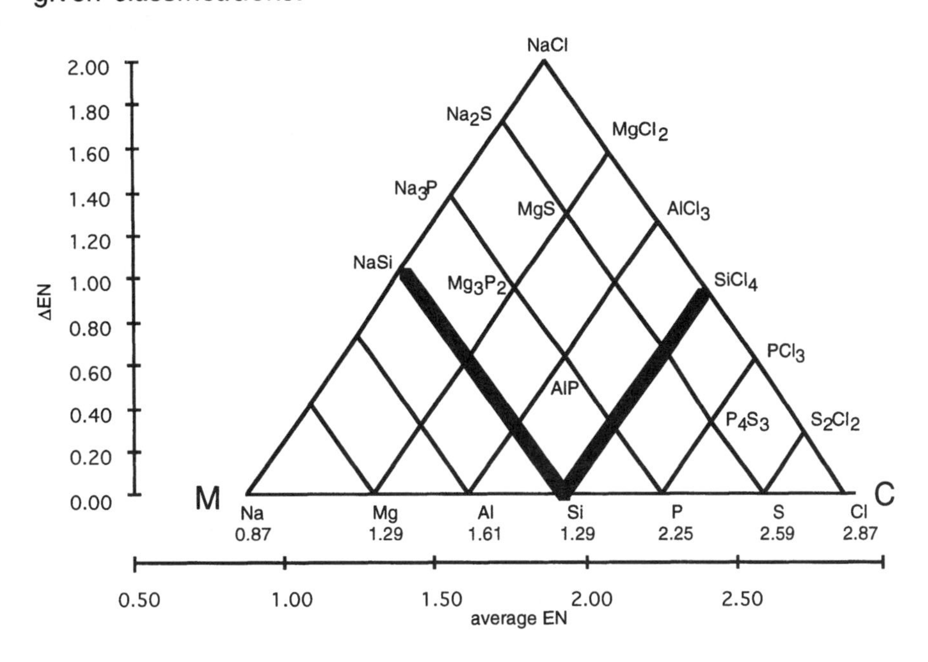

Binary Compounds	Classification	Binary Compounds	Classification
NaCl	Ionic	Mg_3P_2	Metallic
$MgCl_2$	Ionic	Na_2S	Ionic
$AlCl_3$	Covalent	Na_3P	Metallic
$SiCl_4$	Covalent	NaSi	Metallic
PCl_3	Covalent	P_4S_3	Covalent
S_2Cl_2	Covalent	AlP	Metallic
MgS	Ionic		

5-91 No, the compound could be metallic.

5-93

(a) Hg=1.76	S=2.59	EN_{ave}=2.18	ΔEN=0.83	Covalent
(b) Ga=1.76	Sb=1.98	EN_{ave}=1.87	ΔEN=0.22	Semimetal
(c) Li=0.91	N=3.07	EN_{ave}=1.99	ΔEN=2.16	Ionic
(d) Na=0.87	Br=2.69	EN_{ave}=1.78	ΔEN=1.82	Ionic
(e) Sn=1.82	Br=2.69	EN_{ave}=2.26	ΔEN=0.87	Covalent
(f) Na=0.87	P=2.25	EN_{ave}=1.56	ΔEN=1.38	Ionic/Metallic
(g) In=1.66	P=2.25	EN_{ave}=1.96	ΔEN=0.59	Covalent with some semimetal character
(h) In=1.66	N=3.07	EN_{ave}=2.37	ΔEN=1.41	Ionic/Covalent
(i) Te=2.16	O=3.61	EN_{ave}=2.88	ΔEN=1.45	Covalent

5-95

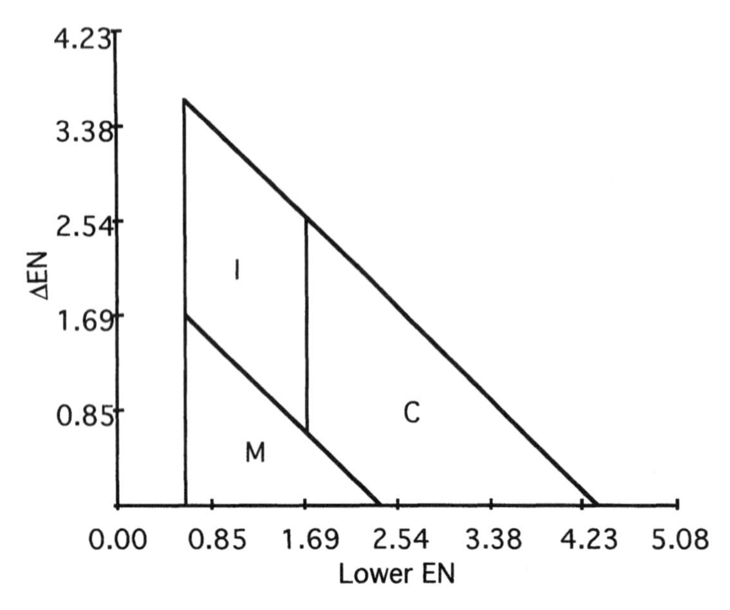

This plot emphasizes the importance of the minimum electronegativity which determines how a bond should be classified. A plot of ΔEN versus lower EN is shown on the left. This plot shows that metallic bonding character increases until the lower EN reaches a particular value at which point a discontinuity in bond type occurs. (Taken from G. Sproul J.Phys.Chem. 93 13221(1994))

5-97 (a) $MgZn_2$ (b) SiO_2 (c) GaAs (d) CsF (e) Hg_2Cl_2 (f) HgO

5-99 NO has a ΔEN of 0.54 and is an insulator. CdLi has a ΔEN of 0.61 and is a conductor.

5-101 P_4O_6 is a covalent molecule since the ΔEN is 1.34 and the average EN is 2.93
Mg_3N_2 is an ionic compound since ΔEN is 1.78 and the average EN is 2.18
Mg_3Sb_2 is a metallic compound since ΔEN is 0.69 and the average EN is1.64

a) This describes the properties of covalent molecule, P_4O_6.
b) This describes the properties of metallic compound, Mg_3Sb_2.
c) This describes the properties of ionic compound, Mg_3N_2.

5-103
$Re_2Cl_8^{2-}$ $2(x) + 8(-1) = -2$, $x = +3$, Re^{3+}
$Cr_2Cl_9^{3-}$ $2(x) + 9(-1) = -3$, $x = +3$, Cr^{3+}
$Mo_2Cl_8^{4-}$ $2(x) + 8(-1) = -4$, $x = +2$, Mo^{2+}

5-105
(a) $LiAlH_4$ $1(+1) + x + 4(-1) = 0$ $x = +3$, Al^{3+}
(b) $Al(H_2O)_6^{3+}$ $x + 6(0) = +3$ $x = +3$, Al^{3+}
(c) $Al(OH)_4^-$ $x + 4(-1) = -1$ $x = +3$, Al^{3+}

5-107 (e), (g)

5-109 (a)-1 (b)-1 (c)0 (d)+1 (e)+5 (f)+5 (g)+7 (h)+7
Iodine has oxidation number of -1 in HI and KI.
Iodine has oxidation number of +5 in KIO_3 and I_2O_5.
Iodine has oxidation number of +7 in KIO_4 and H_5IO_6.

5-111 Barium has an oxidation number of +2, assuming that the oxygen is -2.

5-113 (a)0 (b)-2 (c) -2 (d)+4 (e)+6 (f)+4 (g)+6 (h)+4
(i) +6 (j) +4 (k) +6

Sulfur has an oxidation number of -2 in H_2S and ZnS.
Sulfur has an oxidation number of +4 in SF_4, SO_2, SO_3^{2-}, and H_2SO_3.
Sulfur has an oxidation number of +6 in SF_6, SO_3, SO_4^{2-}, and H_2SO_4.
Sulfur shows even oxidation numbers from -6 to +6.

5-115 +2) TiO
+3) Ti_2O_3, Ti_2S_3, $TiCl_3$
+4) TiO_2, $TiCl_4$, K_2TiO_3, H_2TiCl_6, $Ti(SO_4)_2$
Titanium has oxidation numbers from +2 to +4.

5-117 (a) C_{CH3} =-3 $C_{central}$ = +1 H = +1 Br = -1
(b) C = -2 H = +1 O = -2
(c) C_{CH3}= -3 $C_{central}$ = +2 H = +1 O = -2
(d) C_{CH3} = -3 $C_{central}$ = 0 H = +1 O = -2

5-119 (a) C_{CH3} = -3 $C_{central}$ = +3 H = +1 O = -2 Cl = -1
(b) C = -2 H = +1 N = -3
(c) C_{CH3} = -3 $C_{C=O}$ = +3 C_{CH2} = -1 O = -2 H = +1
(d) C = -3 H = +1
(e) C_{CH2} = -2 C_{CH} = -1 C_{CH3} = -3 H = +1
(f) C = -1 H = +1

5-121 The partial charge on the H in HCl is 0.11. The partial charge on the Cl is –0.11. However the Lewis structure for HCl indicates that each of the atoms is assigned a formal charge of zero. However in assigning oxidation numbers to HCl the shared electrons are assigned to the more electronegative element, which would be the Cl. This would then have an oxidation number of –1 and the hydrogen has an oxidation number of +1.

5-123 (a) This is an oxidation-reduction reaction. Magnesium is oxidized from 0 in Mg(s) to +2 in $MgCl_2$. Hydrogen is reduced from +1 in HCl to 0 in H_2.
(b) This is an oxidation-reduction reaction. Iodine is oxidized from 0 in I_2 to +3 in ICl_3. Chlorine is reduced from 0 in Cl_2 to -1 in ICl_3.
(c) This is an acid-base reaction.
(d) This is an oxidation-reduction reaction. Sodium is oxidized from 0 in Na(s) to +1 in NaOH. Hydrogen is reduced from +1 in H_2O to 0 in H_2.

5-125 In the reaction 4 C(s) + S_8(l) $\rightarrow$ 4 CS_2(l), the oxidation state of carbon increases from 0 in C(s) to +4 in CS_2(l). Carbon is therefore oxidized. The sulfur changes from an oxidation state of 0 in S_8 to -2 in CS_2. Sulfur is therefore reduced.

5-127 (a) The name phosphorus pentoxide for P_2O_5 does not indicate the number of phosphorus atoms present. Diphosphorus pentoxide would be a better name.
(b) The name iron oxide for Fe_2O_3 does not indicate the charge on the iron. The charge needs to be specified so that you can distinguish whether you are discussing FeO or Fe_2O_3. A better name is iron (III) oxide.
(c) The name chlorine monoxide for Cl_2O does not indicate the number of chlorine atoms. A better name is dichlorine monoxide.
(d) The name copper bromide for $CuBr_2$ does not indicate the oxidation number of the copper atom. A better name is copper(II) bromide.

5-129 (a) P_4S_3 (b) SiO_2 (c) CS_2 (d) CCl_4 (e) PF_5

5-131 (a) $SnCl_2$ (b) $Hg(NO_3)_2$ (c) SnS_2 (d) Cr_2O_3 (e) Fe_3P_2

5-133 (a) $Co(NO_3)_3$ (b) $Fe_2(SO_4)_3$ (c) $AuCl_3$ (d) MnO_2 (e) WCl_6

5-135 (a) aluminum chloride
(b) sodium nitride
(c) calcium phosphide
(d) lithium sulfide
(e) magnesium oxide

5-137 (a) antimony(III) sulfide
(b) tin(II) chloride
(c) sulfur tetrafluoride
(d) strontium bromide
(e) silicon tetrachloride

5-139 (a) H_2CO_3
(b) HCN
(c) H_3BO_3
(d) H_3PO_3
(e) HNO_2

5-141 thiosulfate ion, SSO_3^{2-} or $S_2O_3^{2-}$

5-143 (a) calcite, calcium carbonate
(b) barite, barium sulfate

5-145

Compound	ΔEN	Avg EN	Bond Triangle Classification
FeO	1.94	2.64	Covalent
Fe_2O_3	1.94	2.64	Covalent
$FeCl_2$	1.20	2.27	Covalent
$FeCl_3$	1.20	2.27	Covalent

All of the above compounds are classified as covalent according to bond type. They all lie, however, close to the dividing line between ionic and covalent bonding.

5-147

	I	Br
Partial Charge	+0.065	-0.065
Formal Charge	0	0
Oxidation Number	+1	- 1

In the partial charge calculation, the more electronegative(EN) atom, Br, is partially negative and the less EN atom, I, is partially positive. Remember EN decreases down a column. When assigning oxidation numbers, the less EN atom is given a positive oxidation number. Since IBr is neutral, the oxidation numbers must add to zero. Thus, Br has an oxidation number of -1 and I of +1. The formal charge for each atom is zero. The partial charge representation is the most accurate representation. The formal charge calculation is useful when drawing Lewis structures. The sum of the formal charges must equal the overall charge of the molecule or ion. Oxidation numbers are useful when describing oxidation-reduction reactions.

5-149 There is an ionic bond between the Ba^{2+} two NO_3^- ions. Within the nitrate ion there are covalent bonds between N and O. This compound is different because it contains the polyatomic ion NO_3^-. The Lewis structure for $Ba(NO_3)_2$ is shown below.

Ba^{2+}

5-151 (a) C_3H_6

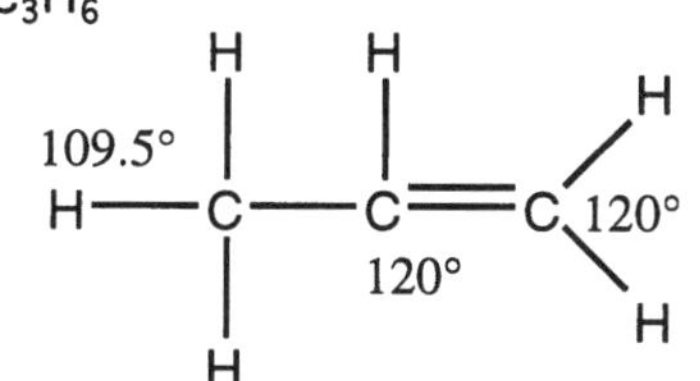

C_3H_3N

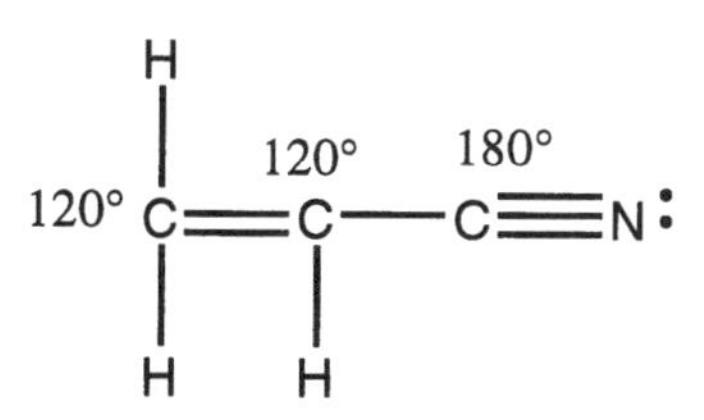

(b) (i) False, the nitrogen in the NO is reduced from a state of +2 to 0. The oxygen doesn't change oxidation state.

(ii) False, the carbon on the C_3H_6 is oxidized from a −3 to a +3 state.

(iii) False, in an oxidation-reduction reaction, at least one of the reactants must be oxidized and another must be reduced. Both reactants cannot be reduced. See the answers for (i) and (ii) for further explanations

(iv) True.

5-153 (a) $2\ PbS + 3\ O_2 \rightarrow 2\ PbO + 2\ SO_2$.

(b) It is an oxidation-reduction reaction. The O_2 is reduced and the S is oxidized.

5-155 MgO.

Chapter 6
Gases

6-1 Metals are good conductors of heat for the same reason that they conduct electricity well. Because of its covalent structure, wood is a poor conductor of heat. When both materials are at the same temperature, below skin temperature, the metal will feel colder because it is carrying the heat away from your skin faster than the wood.

6-3 The **temperature** of a system is a measure of the average kinetic energy of all the particles in the system.

6-5 Water freezes at 32°F, 0°C or 273.15K.

6-7 Kinetic energy can be thought of as the energy of motion. The faster something is moving, the greater its kinetic energy.

6-9 Gases, liquids and solids.

6-11 If both states of matter are present at the same temperature they must both have the same average kinetic energies.

6-13 He, Ne, Ar, Kr, Xe are the atomic elements in the gas phase at room temperature. H_2, N_2, O_2, F_2, Cl_2 are elemental forms of the materials which are also gases at room temperature.

6-15 Low atomic weight elements which are nonmetals are typically gases. Also low molecular weight covalent molecules are commonly gases.

6-17 A gas will always expand to fill up all the volume of its container.

6-19 The volume of a gas is one of the properties described by kinetic molecular theory which is independent of size of the individual gas molecules.

6-21 $\text{Pressure} = \dfrac{\text{Force}}{\text{Area}}$

6-23 The marbles will make contact at a very small surface area. So even with a small force, the pressure is large because the area is very small.

6-25 The height of a column of mercury supported by one atmosphere in a tube twice as large would be 760 mm. Pressure is an intensive property. As the cross-section area of the mercury column increases, the force exerted by the column of mercury increases, force/area remains unchanged.

6-27 $$200\ \text{kPa} \times \frac{1000\ \text{Pa}}{1\ \text{kPa}} \times \frac{1\ \text{atm}}{1.013 \times 10^5\ \text{Pa}} \times \frac{14.7\ \text{psi}}{1\ \text{atm}} = 29.0\ \text{psi}$$

6-29 $$14.7\frac{\text{lb}}{\text{in}^2} \times \frac{1\ \text{in}^2}{(2.54\ \text{cm})^2} \times \frac{(100\ \text{cm})^2}{1\ \text{m}^2} \times \frac{(1000\ \text{m})^2}{1\ \text{km}^2} \times 5.1 \times 10^8\ \text{km}^2 = 1.2 \times 10^{19}\ \text{lb}$$

6-31 $P_1V_1 = P_2V_2$ T=constant

P_1=741.3 mm Hg V_1=0.357 L

P_2=758.1 mm Hg V_2=?

$$V_2 = \frac{P_1V_1}{P_2} = \frac{741.3 \text{ mm Hg} \times 0.357 \text{ L}}{758.1 \text{ mm Hg}} = 0.349 \text{ L}$$

The volume of the balloon will decrease.

6-33 $P_1V_1 = P_2V_2$ T=constant

P_1=200 atm V_1=2.50 L

P_2=1.00 atm V_2=?

$$V_2 = \frac{P_1V_1}{P_2} = \frac{200 \text{ atm} \times 2.50 \text{ L}}{1.00 \text{ atm}} = 500 \text{ L}$$

6-35 $\frac{P_1}{P_2} = \frac{T_1}{T_2}$

P_1=32 psi P_2=60 psi

T_1=21°C=294 K T_2=?

$$T_2 = \frac{T_1P_2}{P_1} = \frac{294 \text{ K} \times 60 \text{ psi}}{32 \text{ psi}} = 5.5 \times 10^2 \text{ K} = 2.8 \times 10^2 \text{ °C}$$

6-37 $\frac{V_1}{V_2} = \frac{T_1}{T_2}$ % change in volume=$\frac{V_2 - V_1}{V_1} \times 100$

T_1=22°C=295 K T_2=75°C=348 K

$$V_2 = \frac{V_1T_2}{T_1} \quad \text{\% change} = \frac{V_1\left(\frac{T_2}{T_1}\right) - V_1}{V_1} \times 100 = \frac{V_1\left(\frac{T_2}{T_1} - 1\right)}{V_1} \times 100$$

$$\text{\% change} = \left(\frac{T_2}{T_1} - 1\right) \times 100 = \left(\frac{348}{295} - 1\right) \times 100 = 17.9\%$$

6-39 The balloons will have an equal contraction in volume upon cooling. Both balloons will start out with the same volume since they have the same number of molecules. As they are cooled they will both obey Charles' Law. Thus they will have the same volume after cooling.

6-41 $2\ NH_3(g) \rightarrow N_2(g) + 3\ H_2(g)$

$$1.38 \text{ L } NH_3 \times \frac{1 \text{ mol } N_2}{2 \text{ mol } NH_3} = 0.690 \text{ L } N_2$$

$$1.38 \text{ L } NH_3 \times \frac{3\ H_2}{2\ NH_3} = 2.07 \text{ L } H_2 \qquad \frac{\text{volume } H_2}{\text{volume } N_2} = \frac{2.07 \text{ L}}{0.69 \text{ L}} = 3:1$$

6-43 $2\ C_2H_2(g) + 5\ O_2(g) \rightarrow 4\ CO_2(g) + 2\ H_2O(g)$

15.0 L of C_2H_2 require 37.5 L O_2 to react. Therefore, O_2 is the limiting reagent.

$$15.0\text{L } O_2 \times \frac{4 \text{ mol } CO_2}{5 \text{ mol } O_2} + 15.0\text{L } O_2 \times \frac{2 \text{ mol } H_2O}{5 \text{ mol } O_2} = 12.0\text{L } CO_2 + 6.0\text{L } H_2O = 18.0 \text{ L Tot Vol Prods}$$

6-45 At the same temperature and pressure equal volumes of gases will contain equal moles of gas. If n is the same for dry air and water vapor, then dry air will weigh more at 29.0 g/mol than water at 18.0 g/mol.

6-47 Look at the ratio of the volumes of the gases:

2.36 L Nitrous oxide(g)→2.36 LN_2(g)+1.18 L O_2(g)

or 2 Nitrous oxide → 2 N_2+1 O_2

The formula must be N_2O.

6-49 PV=nRT

(a)

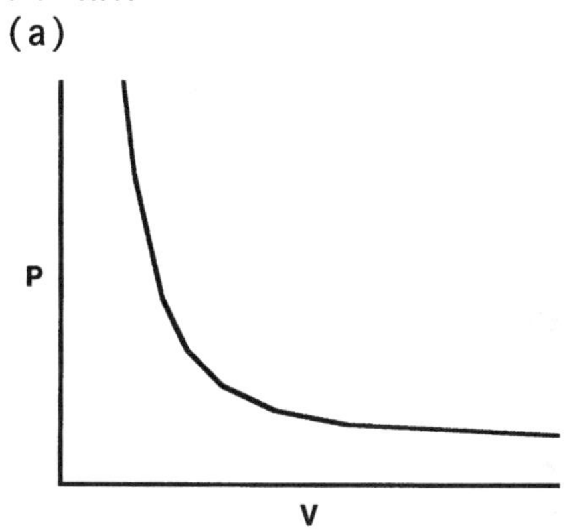

(b)

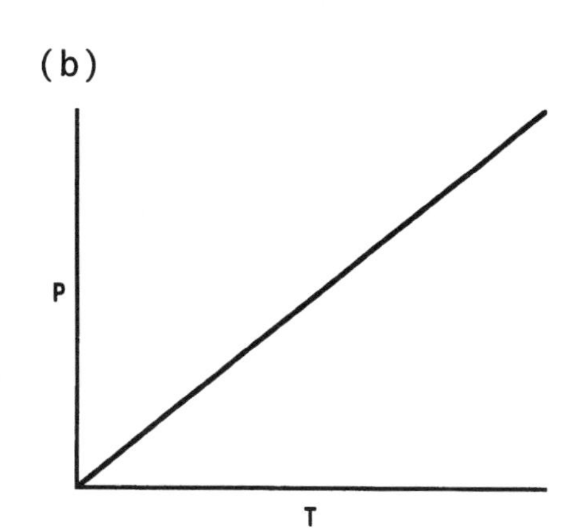

(c)

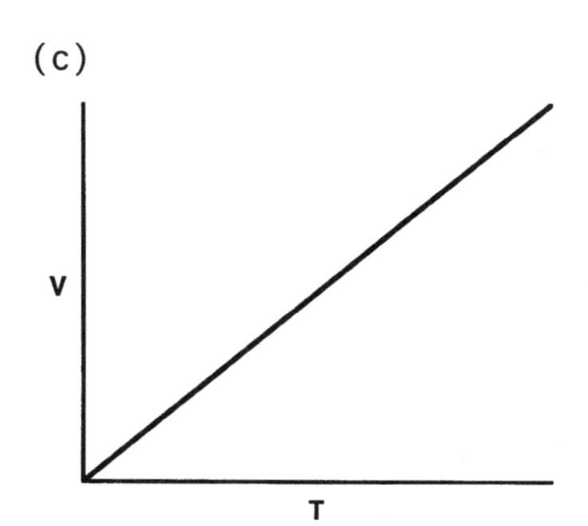

(d)

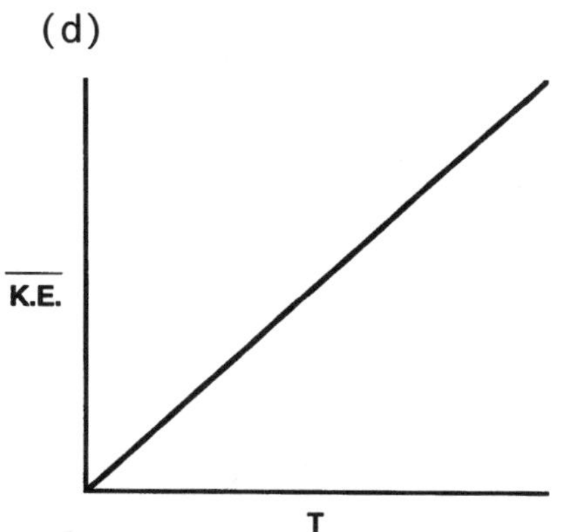

(e)

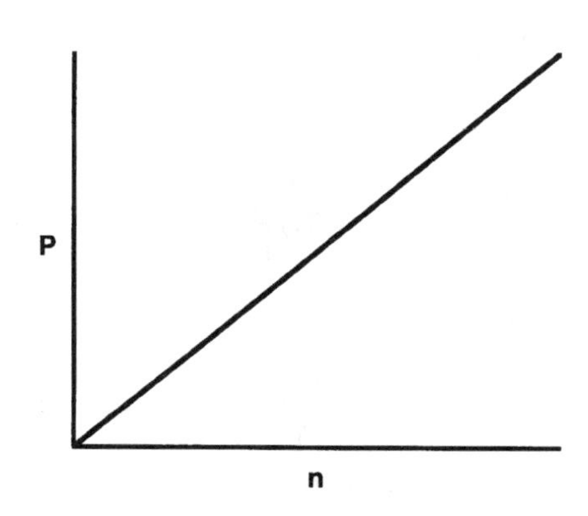

6-51 (a) P=constant $\frac{V}{T}=nR$, n will increase only if V increases more than T. This statement is not always true for an ideal gas.

(b) $\frac{P_1V_1}{T_1}=\frac{P_2V_2}{T_2}$ $\quad V_1\left(\frac{P_1}{P_2}\right)\left(\frac{T_2}{T_1}\right) = V_2$ where $P_2 > P_1$ and $T_2 < T_1$

$\frac{P_1}{P_2}$ is less than 1 and $\frac{T_2}{T_1}$ is less than 1, therefore V_2 must always be less than V_1. The statement is always true for an ideal gas.

(c) $\frac{P_1V_1}{n_1}=\frac{P_2V_2}{n_2}$ $\quad P_1\left(\frac{V_1n_2}{V_2n_1}\right)\left(\frac{T_2}{T_1}\right) = P_2$ where $V_1 > V_2$ and $n_1 > n_2$

$\frac{V_1}{V_2}$ is greater than 1, $\frac{n_2}{n_1}$ is less than 1

Only if the change in V were greater than the corresponding change in n would the pressure need to decrease. The statement is not always true for an ideal gas.

6-53 $$\frac{\$0.50}{100\ ft^3}\times\frac{1\ ft^3}{28.316\ L}\times\frac{22.4\ L}{1\ mol\ N_2}\times\frac{1\ mol\ N_2}{28.014\ g}=\frac{\$0.00014}{g\ N_2}$$

6-55 PV=nRT

$$T=\frac{PV}{nR}=\frac{740\ mm\ Hg\times\frac{1\ atm}{760\ mm\ Hg}\times 1\ L}{1.50\ g\ O_2\times\frac{1\ mol\ O_2}{31.9988\ g}\times 0.08206\frac{L\cdot atm}{mol\cdot K}}=253\ K$$

6-57 PV=nRT $\quad P=\frac{nRT}{V}=\frac{gRT}{MW\ V}=\frac{dRT}{MW}$ where d = density.

$$P=1.118\frac{g}{cm^3}\times 1\ cm^3\times\frac{\frac{1\ mol\ O_2}{31.9988\ g}\times 0.08206\frac{L\cdot atm}{mol\cdot K}\times 273\ K}{0.250\ L}=3.13\ atm$$

6-59 PV=nRT $\quad \frac{n}{V}=\frac{P}{RT}=\frac{1 atm\cdot mol\cdot K}{0.08206\ L\cdot atm\times 313\ K}=0.0389\frac{mol}{L}$

1 mol CH_2Cl_2=84.933 g therefore $\frac{0.0389\ mol\ \times 84.933\frac{g}{mol}}{L}=3.30\frac{g}{L}$ density of CH_2Cl_2 gas.

In the liquid phase, CH_2Cl_2 has a density of $1.336\frac{g}{cm^3}\times\frac{1\ cm^3}{1\ ml}\times\frac{1000\ ml}{1\ L}=1.336\times 10^3\frac{g}{L}$

The density of the liquid is ≅ 400 times greater than the density of the gas.

6-61 1 mol ideal gas occupies 22.4 L at STP

$$\frac{1 \text{ mol He}}{22.4 \text{ L}} = \frac{4.0026 \text{ g}}{22.4 \text{ L}} = 0.179\frac{\text{g}}{\text{L}} = \text{density Helium at STP}$$

The density of air is 1.29 g/L at STP

The lift associated with a balloon is the difference in weight between the displaced gas and that gas doing the displacing.

$$\frac{1.29 \text{ g air} - 0.179 \text{ g He}}{1 \text{ L}} = \frac{1.11 \text{ g}}{\text{L}} \times \frac{28.316 \text{ L}}{\text{ft}^3} \times \frac{1 \text{ lb}}{453.6 \text{ g}} = 0.0694 \frac{\text{lb}}{\text{ft}^3} \text{lift,}$$ Therefore the balloon can lift a weight of 0.07 lbs.

6-63 $3.7493\frac{\text{g}}{\text{L}} \times \frac{22.4 \text{ L}}{\text{mol}} = 84.0\frac{\text{g}}{\text{mol}} = \text{Kr}$; The gas is krypton.

6-65 $\frac{V_1}{T_1} = \frac{V_2}{T_2}$ $\quad V_2 = \frac{V_1 T_2}{T_1} = \frac{(5.0 \text{ L})(78 \text{ K})}{(298 \text{ K})} = 1.3 \text{ L}$

6-67 PV=nRT $\quad P = \frac{nRT}{V} = \frac{4.80 \text{ g } O_3 \times \frac{1 \text{ mol } O_3}{47.997 \text{ g}} \times \frac{0.08206 \text{ L} \cdot \text{atm}}{\text{mol} \cdot \text{K}} \times 298 \text{ K}}{2.46 \text{ L}} = 0.998 \text{ atm}$

$2\ O_3(g) \rightarrow 3\ O_2$

$\frac{P_1 V_1}{n_1} = \frac{P_2 V_2}{n_2}$ $\quad P_2 = \frac{P_1 n_2}{n_1} = \frac{(0.994 \text{ atm})(3 \text{ mol } O_2)}{(2 \text{ mol } O_3)} = 1.49 \text{ atm}$

6-69 $V_2 = \frac{P_1 V_1 T_2}{T_1 P_2} = \frac{(0.9859 \text{ atm})(10.0 \text{ L})(273 \text{ K})}{(393 \text{ K})(1.00 \text{ atm})} = 6.85 \text{ L}$

6-71 The combination of the two initial samples is the same as doubling the amount of one sample.
$P_2 = \frac{P_1 V_1 T_2}{T_1 V_2} = \frac{(0.9859 \text{ atm})\ (20.0 \text{ L})\ (300 \text{ K})}{(393 \text{ K})\ (1.25 \text{ L})} = 12 \text{ atm}$

6-73 P_{He} will be 1/3 of the total pressure=2.5 atm

6-75 $P_{CO_2} = \frac{nRT}{V} = \frac{0.450 \text{ g} \times \frac{1 \text{ mol } CO_2}{44.009 \text{ g}} \times \frac{0.08206 \text{ L} \cdot \text{atm}}{\text{mol} \cdot \text{K}} \times 300 \text{ K}}{1 \text{ L}} = 0.252 \text{ atm}$

$P_{Total} = P_{CO} + P_{CO_2} = 0.200 + 0.252 = 0.452 \text{ atm}$

6-77 The volume of hydrogen does not change when the water vapor is removed. The remaining gas occupies the space available.

6-79 If the gas molecules in a balloon were in a state of constant motion and the motion was not random, the balloon would not appear smoothly rounded. It would be elongated in the direction of the predominating motion and contracted in the other directions.

6-81 The pressure of a gas results from collisions between the gas particles and the walls of the container. Any increase in the number of gas particles in the container increases the number of collisions with the walls and therefore the pressure of the gas.
The average kinetic energy of the gas particles becomes larger as the gas becomes warmer. Since the mass of the gas particles is constant, this means that the average velocity of the particles must increase. The faster the particles are moving when they hit the wall, the greater the force they exert on the wall. Since the force per collision becomes larger as the temperature increases, the pressure of the gas must increase as well.
If we compress a gas without changing its temperature, the average kinetic energy of the gas particles stays the same. There is no change in the speed with which the particles move, but the container is smaller. Thus the particles travel from one end of the container to the other in a shorter period of time. This means that they hit the walls more often. Any increase in the number of collisions with the walls must lead to an increase in the pressure exerted by the gas. Thus the pressure exerted by a gas becomes larger as the volume of the gas becomes smaller.

6-83 (a) $7.0\text{x}10^{-21}$ J/molecule. The temperature determines the average kinetic energy of the molecules. If the NH_3 molecules are at 50°C and have an average kinetic energy of $7.0\text{x}10^{-21}$ J/molecule, then O_2 which is also at 50°C will have the same average kinetic energy.
(b) NH_3. Even though both molecules have the same average kinetic energy, that energy is given by $\frac{1}{2}mv^2$, since a molecule of O_2 is heavier than NH_3 it will have a lower velocity.
(c) No, we would also need the volume and the number of molecules present to determine the pressure PV=nRT.
(d) Since the average kinetic energy is proportional to the temperature, we would need to increase the temperature.
(e) If the temperature is kept the same, then the average kinetic energy is the same, but the pressure would be reduced.
(f) (1) Fix n and V, double the pressure.
(2) Fix T and V, double the number of molecules present.
(3) Fix n and T, cut the volume in half.

6-85 V is a constant so using PV=nRT,

for A $PV = 1.0\,\text{mol} \times \frac{0.08206\ \text{L}\cdot\text{atm}}{\text{mol}\cdot\text{K}} \times 293.15\ \text{K} = 24.1\text{atm}\cdot\text{V}$

for B $PV = 1.0\,\text{mol} \times \frac{0.08206\ \text{L}\cdot\text{atm}}{\text{mol}\cdot\text{K}} \times 308.15\ \text{K} = 25.3\text{atm}\cdot\text{V}$

for C $PV = 2.0\,\text{mol} \times \frac{0.08206\ \text{L}\cdot\text{atm}}{\text{mol}\cdot\text{K}} \times 293.15\ \text{K} = 48.1\text{atm}\cdot\text{V}$

a) From the above calculations, C.
b) Both A and C have the lowest temperature and hence the lowest average kinetic energy.
c) B has a higher temperature therefore a higher average kinetic energy. Since all flasks have atoms of the same mass, the one will the highest kinetic energy will also have the highest velocity.
d) C. Pressure is determined by the number of collisions with the walls. The more collisions the higher the pressure.

6-87 a) i) increase ii) remain the same iii) remain the same iv) decrease
b) remain the same
c) i) increase ii) remain the same (the changes counter balance) iii) decrease
d) The flask with He has more molecules since the pressure is higher while the temperature and volume are the same.

6-89 $Mg(s)+2HCl(aq) \rightarrow Mg^{2+}(aq)+2\,Cl^-(aq)+H_2(g)$

$$\text{mol } H_2 = n = \frac{PV}{RT} = \frac{1.00 \text{ atm} \times 0.500 \text{ L}}{0.08206 \frac{\text{L}\cdot\text{atm}}{\text{mol}\cdot\text{K}} \times 273 \text{ K}} = 0.0223 \text{ mol } H_2$$

$$0.0223 \text{ mol } H_2 \times \frac{1 \text{ mol Mg}}{1 \text{ mol } H_2} \times \frac{24.305 \text{ g Mg}}{1 \text{ mol Mg}} = 0.542 \text{ g Mg}$$

6-91 $CaCO_3(s) \rightarrow CaO(s) + CO_2(g)$

$$150 \text{ kg } CaCO_3 \text{ x } \frac{1000 \text{ g}}{1 \text{ kg}} \text{ x } \frac{1 \text{ mol } CaCO_3}{100.086 \text{ g } CaCO_3} \text{ x } \frac{1 \text{ mol } CO_2}{1 \text{ mol } CaCO_3} = 1.50 \times 10^3 \text{ mol } CO_2$$

$$V = \frac{nRT}{V} = \frac{1.50 \text{ x } 10^3 \text{ x } 0.08206 \frac{\text{L atm}}{\text{mol K}} \text{ x } 296 \text{ K}}{\frac{756}{760} \text{ atm}} = 3.66 \text{ x } 10^4 \text{ L } CO_2$$

6-93 $$10.0 \text{ g } CaCO_3 \times \frac{1 \text{ mol } CaCO_3}{100.086 \text{ g}} \times \frac{1 \text{ mol } CO_2}{1 \text{ mol } CaCO_3} = 0.0999 \text{ mol } CO_2$$

$$V = \frac{nRT}{P} = \frac{0.0999 \text{ mol } CO_2 \times 0.08206 \text{ L}\cdot\text{atm} \times 296 \text{ K}}{0.991 \text{ atm mol}\cdot\text{K}} = 2.45 \text{ L } CO_2$$

6-95 (a), (b), and (d) are correct.

6-97 $$\frac{g}{V} = \text{density} = \frac{P(MW)}{RT},\ MW = \frac{(\text{density})RT}{P} = \frac{5.86 \text{ g x } 0.082056 \text{ L atm x } 273.15 \text{ K}}{1.00 \text{ atm L mol K}}$$

=131 g/mol. The gas is xenon.

6-99 During a rainstorm, gaseous water molecules are condensing to form rain droplets. Thus, the total atmospheric pressure decreases.

6-101 $N_2H_4(g) \rightarrow N_2(g) + 2\,H_2(g)$
From the stoichiometry of the reaction, one mole of gaseous reactant produces three moles of gaseous products. The final pressure will be three times the initial pressure.

6-103 (a) Kinetic energy would remain unchanged, but the frequency of collisions would decrease.
(b) Kinetic energy would decrease as well as the frequency of collisions.
(c) Kinetic energy and frequency of collisions would both increase.
(d) Kinetic energy would remain unchanged, but the frequency of collisions would increase.

6-105 PV=nRT

$$n=\frac{PV}{RT}=\frac{\frac{756}{760}\text{atm}\times 0.2755\text{L}}{0.08206\frac{\text{L}\cdot\text{atm}}{\text{mol}\cdot\text{K}}\times 373\text{ K}}=8.95\times 10^{-3}\text{ mol acetone}$$

$$\text{MW acetone}=\frac{0.520\text{ g}}{8.95\times 10^{-3}\text{ mol}}=58.1\frac{\text{g}}{\text{mol}}$$

6-107 Find empirical formula: $85.63\text{ g C}\times\frac{1\text{ mol C}}{12.011\text{ g}}=7.129\text{ mol C}$

$$14.37\text{ g H}\times\frac{1\text{ mol H}}{1.0079\text{ g}}=14.26\text{ mol H}$$

Empirical formula is CH_2.
$0.45\text{ L }(CH_2)_n$ + excess $O_2 \rightarrow 1.35\text{ L }CO_2 + 1.35\text{ L }H_2O$
at the same T and P, mol amounts are proportional to volumes.
$\text{L }(CH_2)_n$ + excess $O_2 \rightarrow 3\text{ L }CO_2 + 3\text{ L }H_2O$
for which n=3 and the formula for cyclopropane is C_3H_6.

6A-1 The **diffusion** of a gas is the rate at which i mixes with other gases.

6A-3 The speed of a molecule will be greater for a lighter molecule than for a heavier one.

6A-5 $$\frac{\text{Rate }O_2}{\text{Rate }Br_2}=\sqrt{\frac{159.8\text{ g}}{32.00\text{ g}}}=2.236$$

6A-7 The gases will effuse through the pinhole at rates related to the inverse square root of their molar masses. The relative rates are:

N_2	O_2	Ar	CO_2	He
0.189	0.177	0.158	0.151	0.500

6A-9 $$\frac{\text{Time He}}{\text{Time }CH_4}=\frac{6.5\text{ s}}{x}=\sqrt{\frac{4}{16.04}}=\frac{2}{4}=\frac{1}{2},\quad x=13\text{ s}$$

6A-11 $$\frac{\text{rate }N_2O}{\text{rate HCN}}=\sqrt{\frac{\text{MW HCN}}{\text{MW }N_2O}}=\sqrt{\frac{27.026}{44.013}}=0.6220$$

Rate N_2O=(0.6220)Rate NH_3
Total distance traveled=50 Rows

50 rows = Distance N_2O+Distance NH_3
50 rows = Rate N_2O+Rate NH_3
50 rows = 0.6220(Rate NH_3)+Rate NH_3
50 rows = Rate HCN(1.6220)

$$\frac{50\text{ rows}}{1.7837}=\text{Rate HCN}=31\text{ rows}$$

NH_3 travels 31 rows from the back of the room 50-31=19.
The gases will meet on row 19.

6A-13 Since real gas molecules are attracted to each other, they are pulled closer together and their volume will be smaller than that of an ideal gas at the same temperature and pressure.

6A-15 Deviation from ideal gas behavior occurs when attractive forces between gas molecules or molecular volumes become significant. This occurs at low temperatures or high pressures.

6A-17

P(atm)	V(L)	PV(atm·L)
1	1	1
45.8	0.01705	0.7809
84.2	0.00474	0.3991
110.5	0.00411	0.4542
176.0	0.00365	0.6424
282.2	0.00333	0.9397
398.7	0.00313	1.248

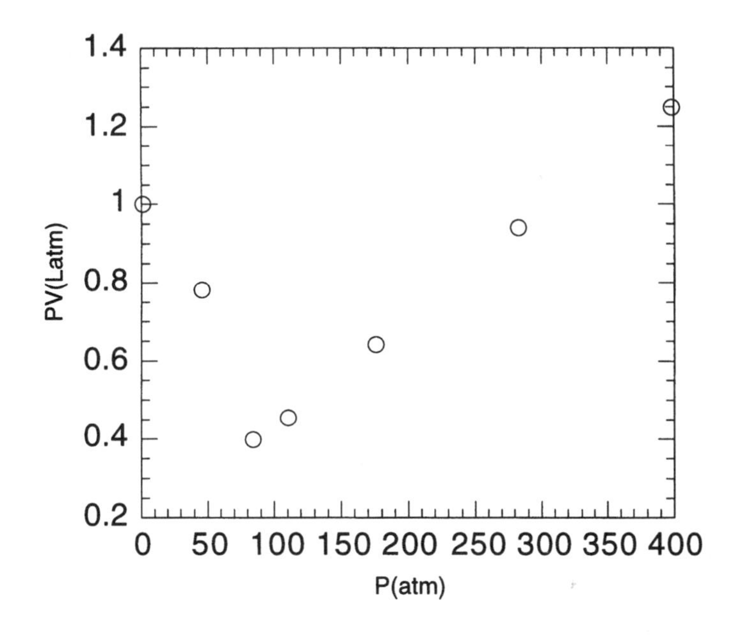

If Boyle's law was followed, PV for acetylene should be constant at a fixed temperature. Clearly, the data presented here show that this is not the case. As the pressure increases, the attractive forces between particles predominates in causing deviation from ideal behavior. As the pressure increases even more, the space occupied by individual gas particles becomes more significant. At the higher pressures, the relative particle size becomes the predominant cause of deviation.

6A-19 The van der Waals constants, b, for the gases are

He 0.02370 L/mol
Ne 0.01709 L/mol
Ar 0.03219 L/mol

The values of b are proportional to the volumes of the atoms. Using these values, we can compare the relative volumes of atoms.

$\frac{He}{Ne} = 1.1$ and $\frac{Ar}{Ne} = 1.2$

6A-21 For CO_2 at 100 atm the negative deviation in the value of PV from the ideal gas law indicates that the $\frac{an^2}{V^2}$ term is important at this pressure. However for H_2 the positive deviation from the ideal gas law indicates that it is the size of the molecules which is important so the nb term is more important.

Chapter 7
Making and Breaking of Bonds

7-1 **Kinetic energy** is energy of motion. Gas molecules in motion are an example of kinetic energy. **Potential energy** is energy that is stored within a system. The energy stored within a chemical bond is an example of potential energy.

7-3 Energy can be transferred from one object to another by mechanical (contact)means. Rotation of a car's tires with respect to a road surface gives an automobile kinetic energy.

7-5 **Thermodynamics** is the study of energy transfer and conversion in a chemical reaction.

7-7 **Temperature** is a quantitative measure of whether an object can be labeled "hot" or "cold." **Heat** is a measure of the energy which causes a change in temperature.

7-9 The ball at the top of the building has a large amount of potential energy. Once the ball is released that potential energy is converted into kinetic energy as the ball moves closer to the ground. When the ball is brought to rest on the pavement, the kinetic energy of the falling object, cannot be just "lost." It is transferred into the motions of the individual atoms and molecules in the ball and the pavement with which it is in contact. By transferring the energy to these motions, the average kinetic energy of these materials is increased, hence a rise in temperature.

7-11 The iron atoms in the warm bar would slow down and the atoms in the cooler bar would speed up until they had the same average kinetic energy. When the average kinetic energies are the same, they have the same temperature.

7-13 As a balloon filled with helium becomes warmer, the particles within become more agitated. This induces a greater pressure on the walls, causing the balloon to expand.

7-15 **Heat** is measured in terms of energy, both kinetic and potential. On the other hand the term **hot** describes whether a system feels warm to the touch or has a high temperature which is only a measure of kinetic energy. It is possible for a system to have a high kinetic energy and low potential energy (gaseous helium at a high temperature). The system may be hot, with a higher temperature, compared to its surroundings (say, a metal container). But since there is little potential energy in the gas, it may not have that much heat.

7-17 The first law of thermodynamics states that energy can be transferred between a system and its surroundings in the form of either heat or work. What the system loses the surroundings must gain and vice versa. The energy of a system is conserved only if no exchange with the surroundings is possible.

7-19 The expansion of a gas pushing a piston from a cylinder is an example of a system doing work on its surroundings. The energy of the system will decrease as it is transferred to the piston. A glass of warm lemonade placed in a refrigerator is an example of a system losing heat to its surroundings. The lemonade is considered to be the system losing heat to the surrounding refrigerator.

7-21 In order to be efficient, machines operate in cycles. A cyclic process requires that ΔE_{total} must be zero because the initial and final states are the same. Thus, $\Delta E_{syst} = -\Delta E_{surr}$ which means that we can only extract as much energy as we put in. In other words, you only get out as much work as heat you put in.

7-23 When an exothermic reaction is run under conditions of constant volume, the internal energy of the system will decrease.

7-25 State functions of a system are often associated with intensive properties of the system. Thus temperature, pressure, volume, enthalpy, and energy of an ideal gas are state functions.

7-27 By definition a property of a system is a state function if it depends only on the state of the system, not on the path used to get to that state.
The state functions are: (a), (b), (c), (d).

7-29 In the reaction

$$2H_2(g) + O_2(g) \rightarrow H_2O(l)$$

Heat energy is required to break the O_2 and H_2 bonds, however energy is released in the formation of H_2O bonds. Since heat is given off, it appears that more energy is released in the bond formation process than is required in the bond breaking process. This heat could be measured using a constant pressure calorimeter.

7-31 Calorimetric data can be used to determine the amount of energy that might be obtained from performing a particular chemical process. For example we can determine how much energy would be available to heat our home by burning natural gas if the enthalpy of combustion of natural gas was known.

7-33 Water has the higher heat capacity. Since the same amount of energy is added to both systems, and the temperature of the water went up less, it requires more energy to raise the temperature of water by one degree than it does to raise the temperature of CCl_4 by one degree.

7-35 q=mass x specific heat x ΔT, $\Delta H° = \frac{-46.22 \text{ kJ}}{1.00 \text{ g Mg}} \times \frac{24.305 \text{ g Mg}}{1 \text{ mol Mg}} = -1.12 \times 10^3 \frac{\text{kJ}}{\text{mol Mg}}$ were transferred to the calorimeter. At a constant pressure $\Delta H = q$ and there were 2.17×10^{-3} moles used so $\Delta H = \frac{1933 \text{ J}}{2.17 \times 10^{-3} \text{ moles}} = 891 \text{kJ/mol}$.

7-37

a) $1 \text{mol } H_2O \times \frac{131.29 \text{ kJ}}{1 \text{ mol } H_2O \text{ reacted}} = 131.29 \text{ kJ}$

b) $2 \text{ mol } H_2O \times \frac{131.29 \text{ kJ}}{1 \text{ mol } H_2O \text{ reacted}} = 262.59 \text{ kJ}$

c) $0.0300 \text{ mol } H_2O \times \frac{131.29 \text{ kJ}}{1 \text{ mol } H_2O \text{ reacted}} = 3.93 \text{ kJ}$

d) $0.0500 \text{ mol C} \times \frac{131.29 \text{ kJ}}{1 \text{ mol C reacted}} = 6.56 \text{ kJ}$

7-39 $\Delta H° = \frac{-46.22 \text{ kJ}}{1.00 \text{ g Mg}} \times \frac{24.305 \text{ g Mg}}{1 \text{ mol Mg}} = -1.12 \times 10^3 \frac{\text{kJ}}{\text{mol Mg}}$

7-41 Since processes (1) and (4) are the same they will have the same reaction enthalpy. Furthermore it doesn't mater whether process (1) is taken directly or if the combination of (2) and (3) are followed. The net enthalpy change will be the same, because the net reaction result is the same. This does illustrate how enthalpy is a state function.

7-43 One must know the initial state of the system (the reactants) and the final state of the system (the products). The enthalpy change is simply the enthalpy of the final state minus the enthalpy of the initial state.

7-45 Thermodynamic data are typically reported at 25°C and 1 bar of pressure (1 bar = 0.9869 atm, and the approximation is often made that 1bar ≈ 1atm, but this should not be used for very precise calculations).

7-47 The ° symbol indicates the standard state. The standard state is defined by a unique set of conditions. There is only one standard state. Therefore, the standard state enthalpy change must be unique, while the enthalpy change for the many other possible states of the system can vary widely.

7-49 In an **exothermic** process energy is given off in the course of the reaction, therefore the bond strengths of the products are greater than for the reactants. In an **endothermic** process energy is absorbed by the system to make the reaction proceed, therefore the reactants have stronger bonds than the products.

7-51 The sign of ΔH indicates which side of the reaction, products or reactants, is favored by bonding forces. If ΔH is positive, energy flows into the system. If ΔH is negative, energy flows out of the system.

7-53 (a) Endothermic. Bonds are broken, but no bonds are formed. ΔH is positive.
(b) Exothermic. Bonds are formed, but no bonds are broken. ΔH is negative.
(c) Exothermic. Bonds are formed, but no bonds are broken. ΔH is negative.

7-55 Yes. The enthalpy is dependent upon the amount of substance formed. To form one mole of $CH_4(g)$ from its atoms, 1662.09 kJ of heat is released. To form two moles, 2 x 1.662.09 = 3324.18 kJ is released.

7-57 Endothermic reactions require an input of heat energy.
(a) Breaking chemical bonds requires input of energy. Endothermic.
(b) Condensation changes in phase are exothermic.

7-59 $\Delta H° = 2849.3 \text{ kJ/mol} - 1062.5\text{kJ/mol} - 1608.531\text{kJ/mol} = 178.3 \frac{\text{kJ}}{\text{mol}_{rxn}}$

The reaction is endothermic.

7-61 $\Delta H° = 962.2 \text{ kJ/mol} - 2 \times 498.340 \text{ kJ/mol} - 3243.3 \text{ kJ/mol} = -1284.4 \frac{\text{kJ}}{\text{mol}_{rxn}}$

7-63 $\Delta H° = 2 \times 5169.38 \text{ kJ/mol} + 13 \times 498.340 \text{ kJ/mol}$
$-8 \times 1608.531 \text{ kJ/mol} - 10 \times 926.29 \text{ kJ/mol} = -5313.97 \frac{\text{kJ}}{\text{mol}_{rxn}}$

7-65 $\Delta H° = 6734.3 \text{ kJ/mol} + 6 \times 970.30 \text{ kJ/mol} - 4 \times 3241.7 \text{ kJ/mol} = -410.7 \frac{\text{kJ}}{\text{mol}_{rxn}}$

7-67 $\Delta H° = 2404.3 \text{ kJ/mol} + 2 \times 326.4 \text{ kJ/mol} - 2 \times 416.3 \text{ kJ/mol} - 3076.0 \text{ kJ/mol}$
$= -851.5 \frac{\text{kJ}}{\text{mol}_{rxn}}$

7-69 $\Delta H° = 4 \times 1152.1$ kJ/mol + 11×498.340 kJ/mol - 2×2404.3 kJ/mol

$$- 8 \times 1073.95 \text{ kJ/mol} = -3310.1 \ \frac{\text{kJ}}{\text{mol}_{rxn}}$$

7-71 For $P_2(g)$ and $P_4(g)$ the average P-P bond enthalpy is found from the enthalpies of atom combination.

For P_2: 485.0 kJ/mol for one P-P bond

For P_4: (assuming a square 4 bond structure)
(1199.65 kJ/mol)/4 = 299.91 kJ/mol P-P bonds

The average P-P bond in P_2 is stronger than that in P_4.

7-73 7-73 From Appendix B.13 and Table 7.4

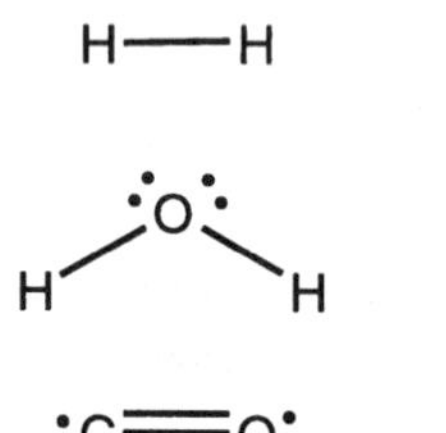

Compound	ΔH_{ac}(kJ/mol$_{rxn}$)
$H_2(g)$	-435.30
$O_2(g)$	-498.340
$H_2O(g)$	-970.30
$CO(g)$	-1076.377
$CH_4(g)$	-1662.09
$CO_2(g)$	-1608.53

Using the structures at the left as a guideline:

C–H bond energy is $\frac{1662.09}{4} = 415.52$ kJ/mol$_{rxn}$

H–H bond energy is 435.30 kJ/mol$_{rxn}$

O–H bond energy is $\frac{970.30}{2} = 485.15$ kJ/mol$_{rxn}$

O=O bond energy is 498.340 kJ/mol$_{rxn}$

C=O bond energy is $\frac{1608.53}{2} = 804.27$ kJ/mol$_{rxn}$

C≡O bond energy is 1076.377 kJ/mol$_{rxn}$

$$\Delta H° = 4 \times 415.52 \text{ kJ/mol} + 2 \times 498.340 \text{ kJ/mol} - 2 \times 804.27 \text{ kJ/mol} - 4 \times 485.15 \text{ kJ/mol} = -890.4 \ \frac{\text{kJ}}{\text{mol}_{rxn}}$$

7-75 $\Delta H° = 2 \times 615.51$ kJ/mol + 3×498.340 kJ/mol - 2×728.18 kJ/mol

$$- 4 \times \frac{1073.95}{2} \text{ kJ/mol} = -878.22 \ \frac{\text{kJ}}{\text{mol}_{rxn}}$$

The bonds of the products are stronger than the bonds in the reactants. This is not surprising since there are six bonds in the products while only five in the reactants.

7-77 a) In order of increasing bond length: H_2, F_2, I_2.

b) Therefore the increasing bond strength would be I_2, F_2, and H_2.

c) Using the enthalpies of atom combination, in order of increasing bond strength, H–H (435 kJ/mol), F–F (158 kJ/mol), I–I (151 kJ/mol). These data support the trend predicted in b).

7-79 This statement can be used when comparing bonds of like or similar atoms, the class of hydrogen halides for example. However, in some cases with multiple bonding or bonds between different classes of atoms, bond length is not an exact predictor of bond strength.

7-81 Take the difference between the two reaction equations and get:
C(graphite) $\rightarrow$ C(diamond)

$$\Delta H^{\circ} = -393.51\frac{kJ}{molCO_2} - \left(-395.41\frac{kJ}{molCO_2}\right) = 1.90\frac{kJ}{mol_{rxn}}$$

7-83

	$\Delta H^{\circ}(kJ/mol_{rxn})$
$NO_2(g) \rightarrow NO(g) + 1/2\ O_2(g)$	+57.1
$1/2\ N_2(g) + O_2(g) \rightarrow NO_2(g)$	+33.2
$1/2\ N_2(g) + 1/2\ O_2(g) \rightarrow NO(g)$	+90.3

The ΔH° for the reaction $N_2(g) + O_2(g) \rightarrow 2NO(g)$ will be
2 x (+90.3 kJ/mol_{rxn}) = +180.6kJ/mol_{rxn}.

7-85

	$\Delta H^{\circ}(kJ/mol_{rxn})$
$C_3H_8(g) \rightarrow 3C(s) + 4H_2(g)$	+103.85
$3C(s) + 3O_2(g) \rightarrow 3CO_2(g)$	-1180.53
$4H_2(g) + 2O_2(g) \rightarrow 4H_2O(g)$	-967.28
$C_3H_8(g) + 5O_2(g) \rightarrow 3CO_2(g) + 4H_2O(g)$	-2043.96

The enthalpy of combustion for propane is -2043.96 kJ/mol_{rxn}.

7-87 The standard enthalpy of formation is defined to be zero for elements in their standard states. These substances are $F_2(g)$ and $P_4(s)$.

7-89 $\Delta H^{\circ} = 135.1\frac{kJ}{mol} - 241.818\frac{kJ}{mol} - \left(-110.525\frac{kJ}{mol} - 46.11\frac{kJ}{mol}\right) = +49.9\frac{kJ}{mol_{rxn}}$

This is the same value as problem 7-60.

7-91 $\Delta H^{\circ} = 2\left(-296.830\frac{kJ}{mol}\right) - 393.509\frac{kJ}{mol} - \left(0 - 89.70\frac{kJ}{mol}\right) = -1076.87\frac{kJ}{mol_{rxn}}$

7-93 $\Delta H^{\circ} = 6\left(-241.818\frac{kJ}{mol}\right) + 4\left(90.25\frac{kJ}{mol}\right) - 4\left(-46.11\frac{kJ}{mol}\right) = -905.47\frac{kJ}{mol_{rxn}}$

7-95 $\Delta H^{\circ} = 2\left(-436.747\frac{kJ}{mol}\right) - 2\left(-391.2\frac{kJ}{mol}\right) = -91.1\frac{kJ}{mol_{rxn}}$

7-97 $100\ ft^3 \times \frac{(0.3048\ m)^3}{1ft^3} \times \frac{(100cm)^3}{1m^3} \times \frac{1mL}{1cm^3} \times \frac{1L}{1000mL} = 2.8\times10^{3}\ L$

$2.83\times10^3\ L \times \frac{1\ mol}{22.4\ L} = 126$ mol at 273K and 1 atm, and the energy released is

$-802.36\ \frac{kJ}{mol\ CH_4} \times (126\ mol\ CH_4) = -1.01 \times 10^5\ kJ$

7-99 ΔH is positive if more enthalpy is required to break the bonds in the reactants than is released when the product bonds are formed.

7-101 In all cases the reaction of interest involves a change from the liquid to the gas phase.

Reaction	ΔH (kJ/mol$_{rxn}$)
$CH_3CO_2H(l) \rightarrow CH_3CO_2H(g)$	52.25
$CH_3CH_2OH(l) \rightarrow CH_3CH_2OH(g)$	42.59
$C_6H_6(l) \rightarrow C_6H_6(g)$	33.96
$CCl_4(l) \rightarrow CCl_4(g)$	32.5

Enthalpies of atom combination were used to calculate the enthalpy change for the reaction and show that intermolecular forces are strongest in $CH_3CO_2H(l)$ and weakest in $CCl_4(l)$.

7-103 The average heat released per mole of covalent bonds broken in the hydrocarbon is 222 kJ.

Compound	$\Delta H_{combustion}$(kJ/mol$_{rxn}$)	Average $\Delta H_{/mol}$ (kJ/mol$_{rxn}$)
$CH_4(g)$	890	223
$C_2H_6(g)$	1560	223
$C_3H_8(g)$	2222	222
$C_4H_{10}(g)$	2877	221
$C_5H_{12}(g)$	3540	221

For each new covalent bond added, the enthalpy of combustion will increase by about 222 kJ/mol$_{rxn}$. Thus, longer chain hydrocarbons release more heat when combusted.

7-105 ΔH for the reaction is found from enthalpies of atom combination to be

ΔH_{cis}=4611.86 kJ/mol +435.30 kJ/mol-5169.38 kJ/mol = -122.2 kJ/mol$_{rxn}$
ΔH_{trans}=4616.58 kJ/mol+435.30 kJ/mol-5169.38 kJ/mol = -117.5 kJ/mol$_{rxn}$

Bonds are stronger in the products.

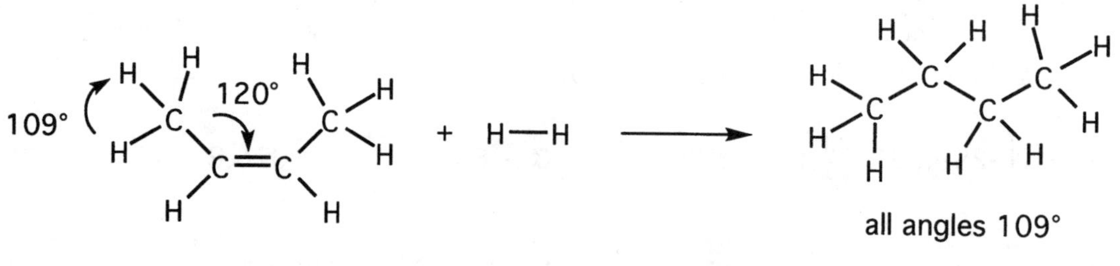

109° 120° + H—H ⟶

all angles 109°

Because the bond strengths are different in the cis and trans forms of 2-butene, a chemical reaction such as the one listed above or a combustion reaction could be used to determine whether the 2-butene was in the cis of trans form. The heat released will be different depending on which isomer is used in the reaction.

7-107 (a) $\Delta H°$ = -1599.3 kJ/mol + 2 x (-192.86 kJ/mol) - (-1272 kJ/mol) - 2 x (-243.358 kJ/mol) = -226 kJ/mol$_{rxn}$
(b) Si-Br: 318 kJ/mol; Si-Cl: 399.8 kJ/mol; Cl-Cl: 243.358 kJ/mol; Br-Br: 192.86 kJ/mol
(c) Si-Cl; smaller size difference between the two bonding atoms. Cl-Cl; smaller atoms can overlap more efficiently; this leads to stronger bonds. Yes, the experimental agree.
(d) Exothermic, because bonding in the products is stronger.

7-109 (d) There are three bonds between the two nitrogens.

7-111 O_2: 498.340/2 = 249.170 kJ/mol; H_2 435.30 kJ/mol; O-H: 463.14 kJ/mol

ΔH = bonds broken in reactants - bonds formed in products

ΔH = 2 x (249.170 kJ/mol) + 2 x (435.30 kJ/mol) - 4 x (463.14 kJ/mol)

= -483.6 kJ/mol

exothermic

7-113 (a) $CH_4(g) + 2\ O_2(g) \longrightarrow CO_2(g) + 2\ H_2O(g)$

(b)

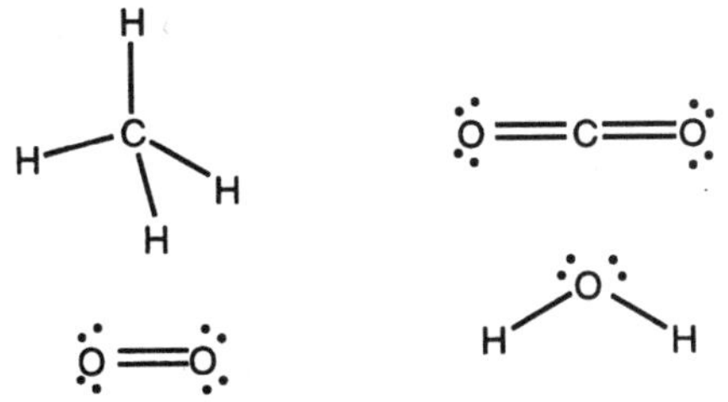

(c) ΔH = 1662 kJ/mol + 2 x 498 kJ/mol - 1608 kJ/mol – 2 x 926 kJ/mol

= -802 kJ/mol

(d) The carbon in methane is oxidized and the O_2 is reduced.

(e) The bonds are stronger in the products

(f) C=O

7-115

Compound	ΔH_{ac}(kJ/mol)
$SiCl_4(g)$	-1599.3
$H_2O(g)$	-926.3
$SiO_2(s)$	-1864.9
$HCl(g)$	-431.6

(a) ΔH = 1599.3 kJ/mol + 2 x 926.3 kJ/mol – 1864.9 kJ/mol – 4 x 431.6 kJ/mol

= –139.4 kJ/mol

(b) Bond strengths are stronger in the products since this is an exothermic rxn.

(c) It would depend on the relative strength differences of the Si–Cl to Si–F bond and the H–Cl to H–F bonds.

(d) The Si-F bond is shorter than Si–Cl, so it should be stronger.

Chapter 8
Liquids and Solutions

8 - 1 On a molecular scale gases are characterized as randomly moving particles widely separated from one another. Solids are characterized as units positioned at relatively short distances in an ordered arrangement. Liquids are characterized in terms of an intermediate arrangement of building units. There is long range ordering for solids, intermediate range ordering for liquids, and essentially no ordering beyond the molecular boundary for gases. On the macroscopic scale, solids retain their shape. A liquid conforms to the lower contours of its container. Gases expand to fill the entire container. The higher compressibility of gases versus that of liquids and solids is another macroscopic manifestation of the differences between phases of matter.

8 - 3 The molecules within a solid vibrate so rapidly as the solid melts that they begin to explore areas outside the rigid confines of the ordered structure of the solid. The liquid is then defined by molecules which are still connected to each other by intermolecular forces, however, the kinetic energy of the molecules in a liquid is high enough that it is not possible for those molecules to maintain an "ordered" structure. As the liquid is heated to boiling, the last of the strong intermolecular forces are broken as the molecule is moving fast enough to overcome any attractions to other molecules in the liquid. Once in the gas phase, the molecule is relatively free of intermolecular forces until it, by chance, comes in contact with another molecule through a collision.

8 - 5 The level of intermolecular forces that exists between molecules of a substance will determine what phase the substance is found in at room temperature. If the substance has very large intermolecular attractions it will be found in the solid phase at room temperature. If the intermolecular forces are weak, the substance will be a gas at room temperature.

8 - 7 **Dipole-dipole** forces arise when molecules have a partial charge separation as an inherent component of the molecular structure. This type of force is purely electrostatic. The interactions between molecules of HCl, which has a large dipole moment, are primarily dipole-dipole.
Dipole-induced dipole forces result when the charge separation associated with a molecular dipole causes (induces) an instantaneous distortion in the electron charge distribution about a molecule which does not have a dipole by itself. This distortion in turn results in an instantaneous charge separation, which in turn is attracted to the first molecular dipole in the same way a dipole-dipole force arises. An example of this would be the solvation of non-polar CCl_4 by the very polar solvent acetone.
Induced dipole-induced dipole forces (dispersion) result from the instantaneous distortion of a molecular electron charge distribution by interaction with another molecular electron charge distribution. Dispersion forces are responsible for keeping the nonpolar CCl_4 molecules in liquid form at room temperature.
Hydrogen bonding is a specific dipole-dipole interaction and results when hydrogen is bonded to electronegative atoms such as N, O, or F. The electronegative element withdraws a large charge density from the proton, making it a very concentrated positive charge. This concentrated positive charge is then attracted to the electronegative atom on a second molecule. In this way the hydrogen acts as a "bridge" between the two electronegative atoms. The hydrogen atom is covalently bound to the first atom and hydrogen bonded to the second. The most ubiquitous example of hydrogen bonding is water.

8-9 A hydrogen bond will only exist when there is a hydrogen atom which is covalently bound to an electronegative element such as N, O or F.

8-11 If the molecule has a dipole moment the molecule will exhibit dipole-dipole forces. If there is a non-polar part of the molecule, there will also be a dipole-induced dipole forces. Dispersion forces are always present. And if the dipole of the molecule is due to an O–H, N-H or H-F bond, there will be hydrogen bonding present.

8-13 In these organic molecules, the intermolecular forces are due primarily to dispersion interactions and one should expect carbon tetrachloride to have a higher boiling temperature because it has the most polarizable electron cloud. Response (e).

8-15 Propane is a lighter molecular weight compound, so the dispersion forces are less in propane.

8-17 The dangling, drawn out structure of n-pentane allows for a closer approach of molecules, thus increasing the dispersion interactions. The structure of isopentane is more compact and symmetric. Overall the isopentane molecule is more spherical. The distance of closest approach of these molecules is greater than with n-pentane and hence the dispersion interactions are weaker.

8-19 The temperature of a substance is a measure of its average kinetic energy. If two sets of the same molecules are at the same temperature, they will have the same average kinetic energy whether they are in the gas phase or liquid phase. And if the molecules have the same mass then they must have the same velocity regardless of what phase they are in.

8-21 The enthalpy of vaporization is the enthalpy change required to break all the intermolecular forces holding a molecule in the liquid phase, freeing it to become a gas. The enthalpy of fusion is the enthalpy change which accompanies the breaking of some intermolecular bonds when the rigid structure of a solid is transformed to the liquid, where intermolecular forces are still present.

8-23 Vapor pressure increases with increasing temperature.

8-25 The kinetic energy of a gas is the energy it has because of the motion of the gas particles. In a gaseous system at any one moment, some of the gas molecules are moving faster than others. The temperature is proportional to the *average* kinetic energy of the particles in the gas.

8-27 A cloth soaked in water feels cool when it is placed on your forehead because the process of evaporation is endothermic. The heat energy required to cause the evaporation of the water in the cloth comes from your forehead.

8-29 The liquid and vapor phases are at equilibrium when the rate of molecules evaporating is the same as the rate of molecules condensing.

8-31 Even at temperatures below the freezing point of water some (but not many) molecules on the surface of the ice will have enough energy to break free from the attractive forces of the rest of the solid. This process is what is called sublimation. It is much the same as water that evaporates from a liquid even when the liquid is below the boiling point of water.

8-33 The force between the mercury atoms is so much stronger than the force between the mercury and the glass that the mercury forms balls to reduce interaction with glass.

8-35 Car wax has a nonpolar surface to it, while water is a very polar molecule. So, on a freshly waxed surface there is little adhesion between water molecules and the wax molecules, so water beads up because it is more attracted to other water molecules than the surface of the car.

8-37 The molecule with the lowest intermolecular forces will have the lower boiling point and the highest vapor pressure. So in each pair the one will have the lower boiling point and the other will have lowest vapor pressure. The lowest boiling point is listed.
a) $CH_3CH_2CH_3$ b) CH_3CH_3 c) CH_3CH_2OH d) $CH_3CH_2CH_2CH_2CH_3$

8-39 A solid does not always get hotter when heat is added. If the solid is at the melting temperature, then any added heat will go into breaking intermolecular bonds, thus melting the material.

8-41 The melting point of a substance is dependent upon the intermolecular forces of the substance. Materials with low intermolecular forces will have low melting points, while materials with a large amount of intermolecular forces will require more kinetic energy (and thus a higher temperature) to break these bonds.

8-43 Denver is at a higher altitude than Salt Lake City. The atmospheric pressure is less in Denver, and water boils at a lower temperature.

8-45 The vapor pressure of water reaches 50 mm Hg at 38.1 °C.

8-47 Increasing the temperature of a liquid will increase the vapor pressure of the liquid. Response (c).

8-49 With the nozzle up, gaseous butane will rise to the top and escape. Since the butane is under pressure inside the can, some of the butane will liquefy. This liquid will drip out of the nozzle when it is pointed down.

8-51 High pressure and low temperature.

8-53 When the horizontal dotted line intersects a solid line there is a phase change. At a pressure of lower than one atm, the freezing point will be slightly less than at one atm and the boiling point will be less than at one atm. The reduction in boiling point will be larger than the reduction in melting point.

8-55 The boiling and melting temperatures of water are much higher than expected from extrapolation of the corresponding values for the dihydrogen compounds of the other members of Group VIA. The concept of the hydrogen bond was introduced to account for the experimental observations. Hydrogen bonds are strong intermolecular bonds in water that arise because of the force of attraction between positively charged hydrogen atoms on one water molecule and the negatively charged oxygen atom on another water molecule. In order to boil water, these intermolecular forces of attraction must be broken and because these forces are stronger for H_2O than the other compounds, a higher temperature is required for water molecules to escape the liquid.

8-57 The rigid hydrogen bonded network in ice results in large hexagonal "open spaces." In the liquid phase water is still hydrogen bonded, but the interactions are not as structured and molecules will flow inside the "open spaces," decreasing the volume that the same number of molecules will take up.

8-59 It takes a large amount of energy to disrupt the hydrogen bonds in water.

8-61 Anesthetics and olive oil are relatively nonpolar.

8-63 As the length of the nonpolar "tail" increases, the polar head of the molecule becomes less significant in determining the physical properties of the compound. (e) 1 heptanol, should display the greater solubility in a nonpolar solvent such as carbon tetrachloride.

8-65 Nonpolar PCl_5 will dissolve better in a nonpolar solvent. The ionic products will dissolve better in a polar solvent. The CCl_4 will favor the reactant. A polar solvent will favor the products.

8-67 Solubility properties of a solid depend upon an interplay of lattice forces acting to hold the particles together in the solid and solvation forces acting to disperse the solute components throughout the solution. In the case of $BaCl_2(s)$, the enthalpy of solution is exothermic. In the case of AgCl(s), the enthalpy of solution is endothermic.

8-69 Tap water has a limited concentration of ions in solution. The solution of NaCl has a much higher concentration of ions. The electric current needed to light the bulb must pass through the solution. The larger the number of ions (charge carriers) the larger the current which can be transported through the solution.

8-71 Using the solubility rules we find that nitrates are soluble, chlorides are generally soluble, and although not specifically mentioned, acetates (OAc) are also generally soluble. Sulfides are usually insoluble, but BaS is an exception. Carbonates and chromates are usually insoluble. Barium carbonate (c) is insoluble.

8-73 Using the solubility rules we find that nitrates are soluble. Lead nitrate (e) is the single salt of the group which is soluble in water.

8-75 a) $Mg(s) + 2\ HCl(aq) \rightarrow MgCl_2(aq) + H_2(g)$

$Mg(s) + 2\ H^{+}(aq) \rightarrow Mg^{+2}(aq) + H_2(g)$

b) $Na_2CO_3(aq) + Ca(NO_3)_2(aq) \rightarrow 2\ NaNO_3(aq) + CaCO_3(s)$

$CO_3^{-2}(aq) + Ca^{+2}(aq) \rightarrow CaCO_3(s)$

8-77 Hydrocarbons do not possess any hydrophilic groups.

8-79 Solubility NaCl > CH_3COOH > CH_3CH_2OH > $CH_3CH_2CH_2CH_3$
NaCl is an ionic compound which is readily soluble in polar water. The acid group ion CH_3COOH makes this quite soluble, but not as much as an ionic species. The –OH group on CH_3CH_2OH is not as polar as the acid, making this less soluble, but there are no hydrophilic groups on $CH_3CH_2CH_2CH_3$, making it the least soluble.

8-81 The hydrogen bonds between the ethanol molecules require more energy to break than the dipole-dipole forces of the dimethyl ether.

8-83 Salt Lake City. At Denver, also known as the Mile High City, water boils at lower temperature because of the decreased atmospheric pressure. A cup of tea made in Salt Lake City would be hotter.

8-85 Figure (a).

8-87 Dispersion, dipole-dipole, and hydrogen bonding all have to be broken in the solvent and solute. In addition, each of these interactions is present in the final solution. Since methanol is a polar molecule, it should be very soluble in the water.

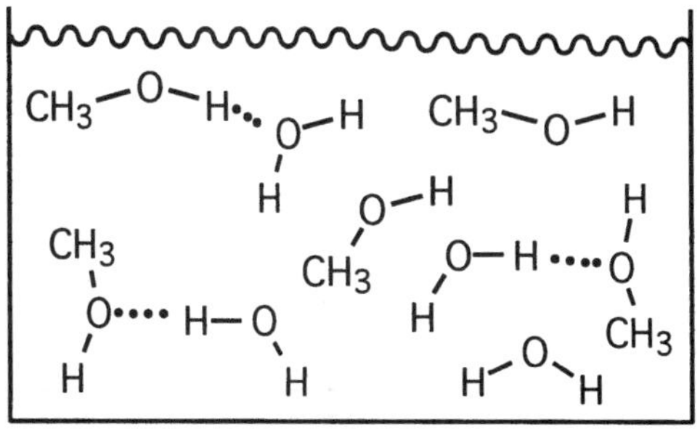

8-89 (a) $Fe(OH)_3$ (b) no precipitate (c) $PbCl_2$

8-91 (a) 1-Pentanol experiences hydrogen bonding and should have a higher boiling point than pentanal and hexane. Butanoic acid experiences strong dipole-dipole interactions in addition to hydrogen bonding. Pentanol should have a lower boiling point than butanoic acid.
(b) The addition of 1C to the chain for each aldehyde listed in the table leads to an increase in the boiling point by ~25°C. Nonal should have a boiling point close to 196°C.

8-93 (a) 1-octanol < dibutyl ether < octane
(b) 1-octanol > dibutyl ether > octane

8-95 At the same temperature, the specific heat of gaseous water should be less than that of liquid water. In the vapor phase, water molecules do not experience as strong of intermolecular forces as in the liquid. Therefore, it would take less heat energy to change the temperature of one gram of gaseous water by one degree Celsius.

8-97 The compound with the higher boiling point must have stronger intermolecular forces. The compound with the lower boiling point has the higher vapor pressure. Both of these compounds probably have a dipole moment because of their relatively high boiling points.

8-99 a) Vapor pressure: $PH_3 > AsH_3 > NH_3 > SbH_3$. The trend for vapor pressure will be opposite that of boiling point.
b) CH_3OH is the best solvent in the group for NH_3 because it will also form hydrogen bonds with the NH_3 solute.

8-101

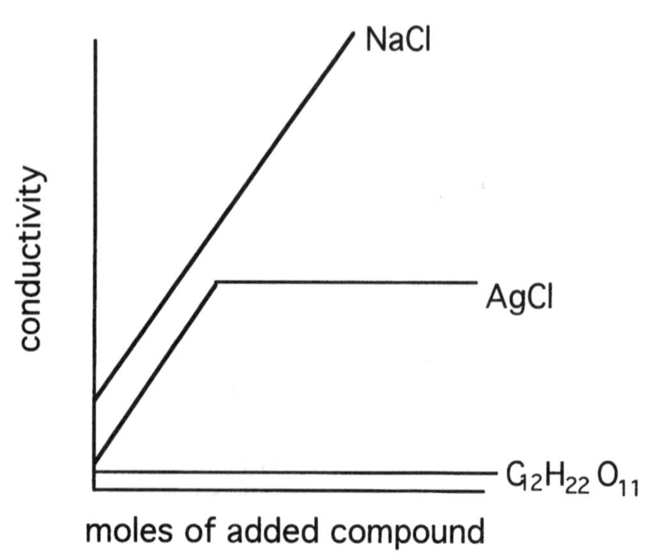

8A-1 **Colligative properties** are properties of solutions that will vary with the concentration of the solute particles in the solution.

8A-3 (a) hexane and heptane would be the closest to an ideal solution.

8A-5 The vapor pressure of a pure liquid depends only upon the temperature. The vapor pressure of a solution, however, is lower than that of the pure liquid by an amount proportional to the added solute.

8A-7 First we need to find the moles of each constituent of the solution:

$$\text{moles}_{C_5H_{12}} = \frac{500\,\text{g}}{72.150\,\frac{\text{g}}{\text{mol}}} = 6.930\,\text{mole}$$

$$\text{moles}_{C_7H_{16}} = \frac{500.\text{g}}{100.203\,\frac{\text{g}}{\text{mol}}} = 4.990\,\text{mole}$$

then

$$\chi_{C_5H_{12}} = \frac{6.930\,\text{mole}}{6.930\,\text{mole} + 4.990\,\text{mole}} = 0.581$$

$$\chi_{C_7H_{16}} = \frac{4.990\,\text{mole}}{6.930\,\text{mole} + 4.990\,\text{mole}} = 0.419$$

and $P_T = \chi_{C_5H_{12}} \times P^{\circ}_{C_5H_{12}} + \chi_{C_7H_{16}} \times P^{\circ}_{C_7H_{16}}$,

$$P_T = 0.581 \times 420\,\text{mm Hg} + 0.491 \times 36.0 = 262\,\text{mm Hg}$$

8A-9 Boiling occurs when the equilibrium vapor pressure of a liquid equals atmospheric pressure. Since a solution has a lower vapor pressure at a given temperature than that of the pure liquid, the solution would need to be heated to a higher temperature to boil.

8A-11 $\Delta T = k_b\,m$. If P_4 is the solute, then using the data from Table 8A.1

$$T - T^{\circ} = k_b \times m,$$

$$T = T^{\circ} + k_b \times m = 46.23^{\circ} + 2.35\,\frac{^{\circ}\text{C}}{\text{m}} \times \frac{10.00\text{g}}{123.896\text{g}/\text{mol}} \times \frac{1}{0.0250\text{g}} = 53.8\,^{\circ}\text{C}$$

8A-13 $\Delta T = k_b$ m. If I_2 is the solute, then using the data from Table 8A.1

$$T - T^o = k_b \times m,$$

$$T = T^o + k_b \times m = 76.75^o + 5.03\ \frac{^oC}{m} \times \frac{3.41g}{253.8g/mol} \times \frac{1}{0.0500g} = 78.1\ ^oC$$

8A-15 A plot of the freezing point of a solution versus molality would yield a straight line with a slope equal to the negative of the freezing depression constant.

8A-17 $\Delta T = -k_f$ m

$$T^o - T = k_f \times m,$$

$$T = T^o - k_f \times m = 0^o - 1.853\ \frac{^oC}{m} \times \frac{1.00g}{212.208g/mol} \times \frac{1}{0.0456L} = -0.191\ ^oC$$

8A-19 The compound that dissolves into the greatest number of ions will produce the largest freezing point depression. Answer (c), $(NH_4)_2SO_4$ is correct.

8A-21 The salt lowers the melting point of ice, and highways can remain free of ice at temperatures lower than the freezing point of pure water.

8A-23 From the previous problem, it was shown that the percent ionization can be related to the van't Hoff factor as follows: $\%\ \text{ionization} = 200\text{x}\frac{i_{observed}}{i_{theoretical}} - 100$

For a 1.33 % ionized CH_3CO_2H solution with $i_{theoretical} = 2$, $i_{observed} = 1.013$.

$$\Delta T_f = i \times k_f \times m = 1.013 \times \left(\frac{1.853^\circ C}{m}\right) \times 0.100\ m = 0.188\ ^\circ C$$

The solution would freeze at -0.188°C.

8A-25

$$\Delta T_b = k_b \times m,\ m = \frac{\Delta T_b}{k_b} = \frac{0.50\ ^\circ C}{0.515\ ^\circ C/m} = 0.97\ m$$

$$\Delta T_f = k_f \times m = \left(\frac{1.853^\circ C}{m}\right) \times 0.97\ m = 1.8^\circ C$$

The solution would freeze at -1.8°C

8A-27 Hard water is water that contains a large amount of Fe^{3+}, Ca^{2+} and Mg^{2+} ions. These ions will combine with a soap and form an insoluble and inactive precipitate. Water can be softened using an ion exchange method which replaces the "hard" ions with Na^+ ions.

Chapter 9
Solids

9-1 **Crystalline** solids are solids where the particles in the solid are ordered in a regular and repeating pattern from one edge of the solid to the other. **Amorphous** solids are solid materials where there is little or no ordered structure. **Polycrystalline** solids contain aspects of the previous two types. In polycrystalline solids there are many small crystalline structures which are arranged in a random fashion.

9-3 Covalent compounds are typically formed by nonmetals. Metallic compounds are formed by metals. Ionic compounds are formed by a bond between a metal and nonmetal.

9-5 Dispersion forces.

9-7 Lattice energy is the energy required to break ionic compounds into gas phase ions.

9-9 Larger lattice energies are associated with the lattice of a given structure type which has the positive and negative ions separated by the shortest distances. See Coulomb's law. LiF will exhibit the largest lattice energy.

9-11 Larger lattice energies are associated with the lattice of a given structure type which has the positive and negative ions separated by the shortest distances. In each sequence the (+2)(-2) separation increases. In the first series, the anion size increases. In the second series, the cation size increases.

9-13 MgO, with the smaller cation, has a higher lattice energy than CaO. It will require more energy to separate the magnesium ion from the oxide ion than what is required to separate the calcium ion from the oxide ion, hence MgO will be less soluble than CaO.

9-15 The melting points and boiling points of ionic compounds will be much higher than those of covalent compounds because the forces holding these materials together are purely ionic and strongly polar. While in covalent molecules, the intermolecular forces discussed in Chapter 8 are responsible for holding these molecules together and they are not as strong as ionic attractions.

9-17 A **metallic bond** is formed between several atoms which all have a very low AVEE (or electronegativity) and share all their valence electrons. Metallic bonds are formed between elements on the left-hand side of the periodic table.

9-19 In molecular, ionic and network covalent solids the electrons are localized either on atoms or in individual bonds. In metallic bonds the electrons are delocalized.

9-21 The delocalization of valence electrons throughout a metal allows us to model the forces which act to make metals malleable and ductile. The arrangement of atoms would suggest that layers could be deformed as the spherical atoms "roll by one another." The delocalized valence electron charge distribution should correspondingly stretch and distort to accommodate the stress being applied. No rupture of bonds is needed to reorient the atom arrangement as long as the metal fabric is not broken.

9-23 The ionic solids as a class are most likely to contain a compound that is a poor conductor of electricity when solid but a very good conductor when molten.

9-25 (a), (d), (e)

9-27 **Semimetals** are elements which have intermediate values for AVEE (or electronegativity). The AVEE is too small to effectively form covalent bonds. However, the AVEE's are usually large enough that they form only partially delocalized metallic bonds, making these elements only semi-conductive.

9-29 Semi-conductors are typically formed by semimetals, which are elements that are found on the "border" between the metals on the left 2/3 of the periodic table and the nonmetals on the right 1/3 of the periodic table.

9-31 a) silicon and carbon
b) carbon and oxygen
c) cadmium and lithium

9-33 The structures of the indicated elements are so designated because the arrangement of atoms in these elements has been found to conform to the indicated models, *simple cubic* for polonium, *body-centered cubic* for iron, and *hexagonal closest packed* for cobalt.

9-35 The larger the number of next nearest neighbors, the larger the number of induced dipole-induced dipole interactions. Both cubic closest packed and hexagonal closest packed structure types allow for twelve next nearest neighbors.

9-37 Hexagonal closest packed, cubic closest packed, body-centered cubic.

9-39 The picture at right shows a 3d cut-away for a simple cubic packing of spheres. The sphere marked with an A is partially hidden in the second layer. It is directly touching a sphere in front and in back. Additionally it touches one above it and one below it. There is also one touching it to the left. If the marked sphere were inside a crystal it would also have contact with one other sphere on the right. This gives it a coordination number of 6.

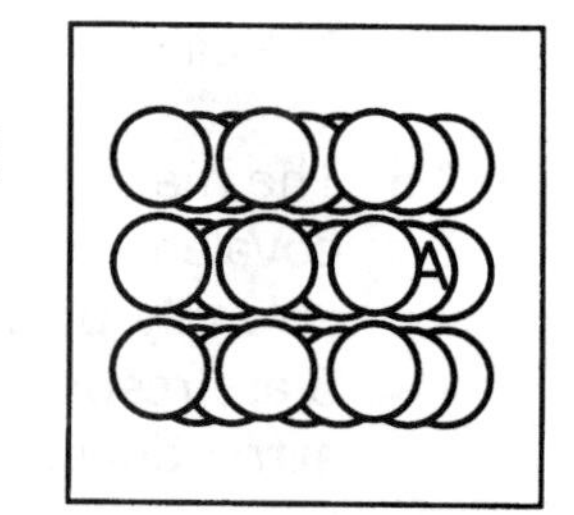

9-41 A metal may want to increase its coordination number to increase the number of other atoms to which it is bound. However, it may also choose another structure to increase the number of atoms which are "nearby" even though they are not in direct contact.

9-43 A **unit cell** is the simplest repeating unit in a crystalline structure.

9-45 Cs metal and CsCl both have a body-centered cubic unit cell. Cs metal crystals are held together by metallic bonds whereas the CsCl crystal is held together by ionic bonds. In each case each atom sits on a lattice point. However in CsCl, the chloride ions define the corners of the unit cell with the Cs^+ in the center, whereas in Cs metal, each atom could define a corner of a unit cell.

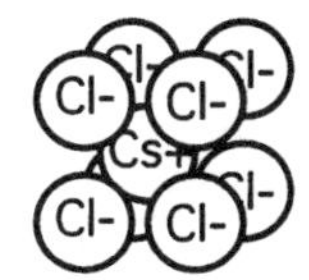

9-47 The density is mass per unit volume. If metal atoms are more closely spaced then the metal will have a higher density since they will take up less volume.

9-49 $$\text{density}_{Ag} = \frac{4 \times \left(\dfrac{107.868 \text{ g Ag}}{1 \text{ mol}} \times \dfrac{1 \text{ mol}}{6.022 \times 10^{23} \text{ atoms}}\right)}{\left(0.40862 \text{ nm} \times \dfrac{1 \text{ m}}{10^9 \text{ nm}} \times \dfrac{100 \text{ cm}}{1 \text{ m}}\right)^3} = 10.50 \text{ g/cm}^3$$

9-51 $$\text{density}_{Ca} = \frac{N \times \left(\dfrac{40.078 \text{ g Ca}}{1 \text{ mol}} \times \dfrac{1 \text{ mol}}{6.022 \times 10^{23} \text{ atoms}}\right)}{\left(0.5582 \text{ nm} \times \dfrac{1 \text{ m}}{10^9 \text{ nm}} \times \dfrac{100 \text{ cm}}{1 \text{ m}}\right)^3} = 1.55 \text{ g/cm}^3$$

N = 4. This suggests face-centered cubic.

9-53 $$\frac{4 \times \left(MW \times \dfrac{1 \text{ mol}}{6.022 \times 10^{23} \text{ atoms}}\right)}{\left(0.3608 \text{ nm} \times \dfrac{1 \text{ m}}{10^9 \text{ nm}} \times \dfrac{100 \text{ cm}}{1 \text{ m}}\right)^3} = 8.95 \text{ g/cm}^3$$

atomic mass = 62.9. The metal is copper.

9-55 $$\frac{N \times \dfrac{25.939 \text{ g LiF}}{1 \text{ mol}} \times \dfrac{1 \text{ mol}}{6.022 \times 10^{23} \text{ atoms}}}{\left(0.4017 \text{ nm} \times \dfrac{1 \text{ m}}{10^9 \text{ nm}} \times \dfrac{100 \text{ cm}}{1 \text{ m}}\right)^3} = 2.640 \text{ g/cm}^3$$

N = 4

9-57 $$V_{unit\ cell} = 2\,\frac{\left(\frac{51.996\text{ g Cr}}{1\text{ mol}} \times \frac{1\text{ mol}}{6.022 \times 10^{23}}\right)}{7.20\ \frac{g}{cm^3}}$$

$$V_{unit\ cell} = a^3 = 2.40 \times 10^{-23}\ cm^3$$

$$a = \sqrt[3]{2.40 \times 10^{-23}\ cm^3} = 2.88 \times 10^{-8}\ cm$$

$$d^2_{body} = \left(\sqrt{2}a\right)^2 + a^2 = 3\,a^2$$

$$d_{body} = \sqrt{3}a = \sqrt{3} \times (2.88 \times 10^{-8}\ cm) = 5.00 \times 10^{-8}\ cm$$

$$d_{body} = 4\ r_{Cr}$$

$$r_{Cr} = \frac{5.00 \times 10^{-8}\ cm}{4} = 1.25 \times 10^{-8}\ cm$$

9-59 From the geometry of the body-centered cubic structure, we find the shortest distance between barium atoms will occur down the body diagonal where

$$d_{body} = \sqrt{3}\ a = 4r_{Ba}$$

$$r_{Ba} = \frac{\sqrt{3}}{4} \times 0.5025\ nm \times \frac{1m}{10^9 nm} \times \frac{100cm}{1m} = 2.176 \times 10^{-8}\ cm$$

The shortest distance between barium atoms will be 2 r_{Ba}.
The Ba-Ba distance = 4.352×10^{-8} cm.

9-61 The unit cell for the cesium chloride structure type places the ($r_{Tl} + r_I$) distance along the body diagonal of the cube with edge distance a. From the geometry of the cubic structure type,

$$2(r_{Tl} + r_I) = \sqrt{3}\ a$$

$$r_{Tl} + r_I = \frac{\sqrt{3}}{2} \times .419.8\ nm = 0.3635\ nm$$

$$r_{Tl} = 0.3635\ nm - r_I = 0.3635 - 0.216 = 0.148\ nm$$

9-63

	$\Delta H_{fus}(kJ/mol_{rxn})$	$\Delta H\ X_2(g) \rightarrow 2X(g)\ (kJ/mol_{rxn})$
F_2	0.51	157.98
Cl_2	6.41	243.358
Br_2	10.8	192.86
I_2	15.3	151.238

The force that holds the atoms together, the covalent bond, is much stronger than the dispersion forces which hold the molecules together.

9-65

$\Delta H_{LE}(BeCl_2) = -\Delta H_{ac} + \Delta H_{IE} + \Delta H_{EA}$

$= 1058.1 + 2656.5 - 2 \times 349$

$= 3017$ kJ/mol$_{rxn}$

$\Delta H_{LE}(BaCl_2) = 2052$ kJ/mol$_{rxn}$ Problem 9-64

The products of charges are the same for $BeCl_2$ and $BaCl_2$ but Be has the smaller radius and $BeCl_2$ has the larger lattice energy.

9-67

A.	(e)	F.	(c)
B.	(f) or (a)	G.	(b)
C.	(d) or (b)	H.	(e)
D.	(f) or (a)	I.	(f),(a), or (c)
E.	(c) or (b)	J.	(d) or (c)

9-69

(a)	molecular	(e)	metallic	(i)	ionic/metallic
(b)	molecular	(f)	ionic	(j)	molecular
(c)	metallic	(g)	metallic	(k)	ionic/metallic
(d)	molecular/ionic	(h)	ionic	(l)	molecular/metallic (semimetal)

Chapter 10
An Introduction to Kinetics and Equilibrium

10-1 Reactions that go to completion leave no unreacted limiting reagent behind. Reactions that go to equilibrium may leave an appreciable amount of unreacted starting materials.

10-3 The symbol [NO] refers to the concentration (moles/liter) of nitric oxide **at equilibrium**. The symbol (NO) refers to the concentration of nitric oxide at any given moment, not necessarily at equilibrium.

10-5 The ratio of the equilibrium concentrations of trans-2-butene to cis-2-butene at 400°C is 1.27 according to the data in section 10.2. This ratio will always have this value no matter how much material is present, so long as the system is at equilibrium at 400°C.

10-7 This is a true statement.

10-9 The rate of a chemical reaction is equivalent to the measured change in amount of a reactant or product divided by the measured time interval over which the change in amount occurred.

10-11 The **rate constant** is the proportionality constant between the rate of the reaction and the concentrations of reactants in the rate law. The rate constant is only one part of the entire rate law.

10-13 Since the rate of the forward reaction is dependent upon the concentrations of the reactants, as those reactants are used up in the reaction, the forward reaction slows down. Conversely as the reaction proceeds there are more products present, and the rate of the reverse reaction will be dependent upon those concentrations. Therefore the rate of the reverse reaction will increase as the reaction proceeds forward.

10-15 Equilibrium was initially defined as when the forward reaction appears to stop and the concentrations of the products and the reactants do not change. However a more correct definition of equilibrium is when the rate of the forward reaction is equal to the rate of the reverse reaction.

10-17 (d) $K_c = \frac{[ClF_3]^2}{[Cl_2]\ [F_2]^3}$

10-19 (a) $K_c = \frac{[OF_2]^2}{[O_2]\ [F_2]^2}$ (b) $K_c = \frac{[SO_3]^2}{[O_2]\ [SO_2]^2}$ (c) $K_c = \frac{[SO_2Cl_2]^2}{[SO_3]^2\ [Cl_2]^2}$

10-21 (a) $K_c = \frac{[CO_2]^2}{[CO]^2\ [O_2]}$ (b) $K_c = \frac{[CO]\ [H_2O]}{[CO_2]\ [H_2]}$ (c) $K_c = \frac{[CH_3OH]}{[CO]\ [H_2]^2}$

10-23 $K_c(c) = K_c(a) \times K_c(b) = \frac{[N_2]\ [O_2]^2}{[NO_2]^2} = \frac{[N_2]\ [O_2]}{[NO]^2} \times \frac{[NO]^2\ [O_2]}{[NO_2]^2}$

$K_c(c) = (4.3 \times 10^{18}) \times (3.4 \times 10^{-7}) = 1.5 \times 10^{12}$

10-25 $K_c = \dfrac{[NO]^2[Cl_2]}{[NOCl]^2} = \dfrac{(0.0222)^2\ (0.0111)}{(0.978)^2} = 5.72 \times 10^{-6}$

10-27 $Q_c < K_c$ The reaction must shift toward the products to reach equilibrium.

10-29 $Q_c < K_c$ Under these conditions there is not enough product formed so the reaction would shift to the right to reach equilibrium. Knowing K_c alone does not provide enough information to say which way the reaction will proceed, concentrations are needed.

10-31 The initial concentration of a reactant, (x), is some starting value either selected by the experimenter or determined through quantitative analysis of the system under study. The concentration of a reactant at equilibrium, [x], is a unique value at a given temperature and depends on the amounts of the other components associated with the reaction system. The change in concentration, $\Delta(x)$, that occurs as the reaction comes to equilibrium is the difference between the two concentrations defined above. $\Delta(x) = (x) - [x]$.

10-33 One change in N_2 concentration corresponds to 3 changes in H_2 concentration.

$$\frac{\Delta(H_2)}{\Delta(N_2)} = \frac{3}{1} \text{ and } 1\ \Delta(H_2) = 3\ \Delta(N_2)$$

10-35

$$2\ NH_3(g) \rightleftarrows N_2(g) + 3\ H_2(g)$$

2 $NH_3(g)$	$\rightleftarrows$	$N_2(g)$	+	3 $H_2(g)$
$-2\ \Delta c$		$+\Delta c$		$+3\Delta c$

$2\ \Delta c = 0.234$
$\Delta c = 0.117$
$\Delta(N_2) = 0.117$
$\Delta(H_2) = 0.351$

10-37 (e) $\Delta(F_2) = 3\ \Delta(Cl_2)$

10-39

	$N_2(g)$	+	3 $H_2(g)$	$\rightleftarrows$	2 $NH_3(g)$
initial	1.000 M		1.000 M		0
change	$-\Delta c$		$-3\Delta c$		$2\Delta c$
equilibrium	0.922 M		1.000-3(0.078)		2(0.078)

$[N_2]_{eq} = 0.922$ M $[H_2]_{eq} = 0.766$ M $[NH_3]_{eq} = 0.156$ M

10-41

	$N_2O_4(g)$	$\rightleftarrows$	$2\ NO_2(g)$	$K_c = 5.8 \times 10^{-5}$
initial	0.100 M		0	
change	Δc		$2\Delta c$	
equilibrium	$0.100 - \Delta c$		$2\Delta c$	

$$\frac{(2\Delta c)^2}{(0.100 - \Delta c)} = 5.8 \times 10^{-5}$$

Assume $\Delta c << 0.100$

$4\ \Delta c^2 = 5.8 \times 10^{-6}$

$\Delta c = 1.2 \times 10^{-3}$

Check, $\frac{1.20 \times 10^{-3}}{0.100} \times 100\% = 1.2\%$, the assumption is valid.

$[N_2O_4]_{eq} = 0.099$ M $[NO_2]_{eq} = 2.4 \times 10^{-3}$ M

Check, $\frac{(2.4 \times 10^{-3})^2}{0.099} = 5.8 \times 10^{-5}$, which agrees with the given value.

10-43

	$N_2(g)$	$+\ 3\ H_2(g)$	$\rightleftarrows$	$2\ NH_3(g)$	$K_c = 0.040$ at 500°C
initial	0.10	0.10		0.10	

$Q_c = 1.00 > K_c$, reaction proceeds to the left :

change	$+\Delta c$	$+3\Delta c$		$-2\Delta c$	
equilibrium	$0.10 + \Delta c$	$0.10 + 3\Delta c$		$0.10 - 2\Delta c$	

$$K_c = \frac{(0.10 - 2\Delta c)^2}{(0.10 + \Delta c)(0.10 + 3\Delta c)^3} = 0.040$$

Δc is not small compared to 0.10. So we solve the problem by successive approximation. We find Δc˜0.045 and

$[NH_3] = 0.01$ M $[N_2] = 0.15$ M $[H_2] = 0.24$ M

10-45 $[CO_2] = [H_2] = 1.00$ M at equilibrium, then

$$\frac{1.00^2}{(x - 1.00)^2} = 0.72$$

$$\frac{1.00}{(x - 1.00)} = \sqrt{0.72} = 0.8485$$

$x = 2.18$

The initial concentration of CO and H_2O is 2.18 M.

10-47 MW SO_2Cl_2 = 32.066 + 2(15.999) + 2(35.453) = 134.970 g/mol

$[SO_2Cl_2]$ = 6.75g/(134.970 g/mol) = 0.0500M

	$SO_2Cl_2(g)$	$\rightleftarrows$	$SO_2(g)$	+	$Cl_2(g)$	$K_c = 1.4 \times 10^{-5}$
initial	0.0500 M		0		0	
change	$-\Delta c$		Δc		Δc	
equil	$0.0500-\Delta c$		Δc		Δc	

$$\frac{(\Delta c)\,(\Delta c)}{(0.0500 - \Delta c)} = 1.4 \times 10^{-5}$$

Assume $\Delta c << 0.0500$

$\Delta c^2 = 7.0 \times 10^{-7}$

$\Delta c = 8.4 \times 10^{-4}$

check, $\frac{8.4 \times 10^{-4}}{0.0500} \times 100\% = 1.7\%$, the assumption is valid.

$[SO_2Cl_2]_{eq} = 0.049$ M $[SO_2]_{eq} = [Cl_2]_{eq} = 8.4 \times 10^{-4}$ M

Check $\frac{(8.4 \times 10^{-4})^2}{(0.049)} = 1.4 \times 10^{-5}$, which agrees with the given value.

10-49

	$PCl_5(g)$	$\rightleftarrows$	$PCl_3(g)$	+	$Cl_2(g)$	K_c = 0.0013 at 450K
initial	1.00 M		0		0	
change	$-\Delta c$		Δc		Δc	
equilibrium	$1.00-\Delta c$		Δc		Δc	

$$\frac{\Delta c^2}{1.00 - \Delta c} = 0.0013$$

Assume $\Delta c << 1.00$

$\Delta c^2 = 0.0013$

$\Delta c = 0.036$

Check Δc is 3.6% of the initial concentration. The assumption that $\Delta c << 1.00$ is valid.

$[PCl_5]_{eq}$ = 0.96 M

$[PCl_3]_{eq} = [Cl_2]_{eq}$ = 0.036 M

% decomposition of $PCl_5 = \frac{0.036}{1.00} \times 100\% = 3.6\%$

The difference in the % decomposition of PCl_5 in this reaction carried out at 450 K and the same reaction carried out at 250 K is due to the different values of K_c at the different temperatures. There is less tendency for product formation in this reaction system at the higher temperature.

10-51

	$2\ NO_2(g)$	$\rightleftarrows 2\ NO(g)$ +	$O_2(g)$	$K_c = 3.4 \times 10^{-7}$ at 200K
initial	0.100 M	0	0	
change	$-2\Delta c$	$2\Delta c$	Δc	
equil	$0.100-2\Delta c$	$2\Delta c$	Δc	

$$\frac{(2\Delta c)^2(\Delta c)}{(0.100-2\Delta c)^2} = 3.4 \times 10^{-7}$$

Assume $2\ \Delta c << 0.100$

$$\frac{4\Delta c^3}{(0.100)^2} = = 3.4 \times 10^{-7}$$

$$\Delta c^3 = \frac{3.4 \times 10^{-9}}{4} = 8.5 \times 10^{-10}$$

$\Delta c = 9.5 \times 10^{-4}$

Check, $\frac{2\left(9.5 \times 10^{-4}\right)}{0.100} \times 100\% = 1.9\%$, the assumption is valid.

$[NO_2]_{eq} = 0.098$ M $[NO]_{eq} = 0.0019$ M $[O_2]_{eq} = 9.5 \times 10^{-4}$ M

10-53

	$2\ SO_3(g)$	$\rightleftarrows 2\ SO_2(g)$ +	$O_2(g)$	$K_c = 1.6 \times 10^{-10}$ at 300°C
initial	0.400 M	0	0 M	
change	$-2\Delta c$	$2\Delta c$	Δc	
equil	$0.400-2\Delta c$	$2\Delta c$	Δc	

$$\frac{(2\Delta c)^2(\Delta c)}{(0.400-2\Delta c)^2} = 1.6 \times 10^{-10}$$

Assume $2\ \Delta << 0.400$

$$4\ \Delta c^3 = (0.400)^2(1.6 \times 10^{-10})$$

$$\Delta c^3 = 6.4 \times 10^{-12}$$

$$\Delta c = 1.9 \times 10^{-4}$$

Check, $\frac{3.8 \times 10^{-4}}{0.400} \times 100\% = 0.95\%$, the assumption is valid.

$[SO_3]_{eq} \approx 0.400$M $[SO_2]_{eq} = 3.7 \times 10^{-4}$ M $[O_2]_{eq} = 1.9 \times 10^{-4}$ M

Check, $\frac{(3.7 \times 10^{-4})^2(1.9 \times 10^{-4})}{(0.400)^2} = 1.6 \times 10^{-10}$, which agrees with the given value.

10-55 When the assumption is made that the change in the concentration of a reactant or product is very small, this is interpreted to mean that compared to the initial value the final value will not have changed enough to affect the calculations within experimental error. A comparatively very small value is still a positive number and is not zero. To say that it is zero forces the problem to have no solution.

10-57 Solve the problem neglecting the value of Δc (assumed to have insignificant influence over certain terms). Divide the value of Δc by the smaller of the values relative to which Δ was considered to be negligible. In our test if the ratio is greater than 5% the assumption is considered to be invalid and a more rigorous solution is required.

10-59 Equilibrium constants will change with temperature depending upon the nature of the reaction involved.

10-61 If decreasing the temperature decreases the equilibrium constant, then increasing the temperature should cause an increase in the equilibrium constant, which means the products will be more favored.

10-63 When the pressure changes on a reaction with gaseous components which was initially in a state of equilibrium, the reaction system shifts in the direction which will best offset the externally applied pressure change. A decrease in pressure will favor the shift in reaction direction which generates more gaseous components. An increase in pressure will cause the reaction to shift in the direction which has the smaller amounts of gaseous components.

With an increase in pressure for the reactions listed:
(a) The reaction will shift to the right in the direction of increased production of SO_2Cl_2 and O_2.
(b) The reaction will shift to the right in the direction of forming OF_2.
(c) The reaction will shift to the right in the reaction of formation of NO_2.

10-65 When the concentration of a component of a reaction in equilibrium is changed the reaction system will shift in the direction to reduce the effect of the change in concentration. With an increase in the concentration of the bold component, the reaction will shift in the direction which will decrease the concentration of that component (and all other components on the same side of the reaction equation) while increasing the concentrations of all components on the other side of the reaction equation.
(a) Increasing the concentration of $NO_2(g)$ will shift the reaction to the right until equilibrium is reached.
(b) Increasing the concentration of $O_2(g)$ will shift the reaction to the left until equilibrium is reached.
(c) Increasing the concentration of $PF_5(g)$ will shift the reaction to the right until equilibrium is reached.

10-67 An increase in pressure will force more gas particles to impinge on the water surface. More particles hitting the surface means more are likely to penetrate into the solution. Therefore we should predict that the solubility of a gas in water will increase with an increase in pressure.

10-69 An increase in pressure favors the formation of ammonia in the Haber process because the reaction system is inclined to shift in the direction of fewer numbers of molecules.

10-71 The worst conditions from an equilibrium standpoint are the ones which provide the least amount of product. According to Table 10.3 this would be at high temperature and low pressure.

10-73 $[Ag^+] = 2\,[CrO_4^{2-}]$ $K_{sp} = [Ag^+]^2[CrO_4^{2-}]$

10-75 $K_{sp} = [Sr^{2+}]\,[F^-]^2$

$$\frac{0.107 \text{ g } SrF_2}{1 \text{ L soln}} \times \frac{1 \text{mol } SrF_2}{125.62 \text{ g } SrF_2} = \frac{8.52 \times 10^{-4} \text{ mol } SrF_2}{1 \text{ L soln}}$$

$K_{sp} = (8.52 \times 10^{-4})(2 \times 8.52 \times 10^{-4})^2 = 2.47 \times 10^{-9}$

10-77 $K_{sp} = [Ba^{2+}]\,[SO_4^{2-}]$

Taking the density to be 1g/ml gives : $400{,}000 \text{ g} \times \frac{1\text{ml}}{1\text{g}} = 400{,}000 \text{ ml} = 400 \text{ L}$

$$\frac{1.0 \text{ g}}{400 \text{ L}} \times \frac{1 \text{ mol } BaSO_4}{233.39 \text{ g } BaSO_4} = \frac{1.1\times10^{-5} \text{ mol } BaSO_4}{1\text{Lsoln}} = [Ba^{2+}] = [SO_4^-]$$

$K_{sp} = 1.2 \times 10^{-10}$

10-79 (a) $K_{sp} = [Hg_2^{2+}][S^{2-}] = C_s^2 = 1.0 \times 10^{-47}\ C_s = 3.2 \times 10^{-24}$ mol/L

$$\frac{3.2\times10^{-24} \text{ mol } Hg_2S}{1000 \text{ ml soln}} \times \frac{433.25 \text{ g } Hg_2S}{1 \text{ mol } Hg_2S} = \frac{1.4\times10^{-21} \text{ g } Hg_2S}{1000 \text{ ml soln}} = \frac{1.4\times10^{-22} \text{ g } Hg_2S}{100 \text{ ml soln}}$$

(b) $K_{sp} = [Hg^{2+}][S^{2-}] = C_s^2 = 4 \times 10^{-53}\quad C_s = 6 \times 10^{-27}$ mol/L $= [Hg^{2+}] = [S^{2-}]$

$$\frac{6\times10^{-27} \text{ mol HgS}}{1000 \text{ ml soln}} \times \frac{232.66 \text{ g HgS}}{1 \text{ mol HgS}} = \frac{1.5\times10^{-24} \text{ g HgS}}{1000 \text{ ml soln}} = \frac{2 \times 10^{-25} \text{ g HgS}}{100 \text{ ml soln}}$$

10-81 $K_{sp} = [Ag^+][Br^-] = 5.0 \times 10^{-13}$

In this system the amount of Br^- already in solution will be much greater than the amount that will be added upon dissolving the AgBr so $[Ag^+]=C_s$

$K_{sp} = C_s \times 0.050 = 5.0 \times 10^{-13}$

$C_s = 1.0 \times 10^{-11}$ mol/L

$$\frac{1.0\times10^{-11} \text{ mol AgBr}}{1000 \text{ ml soln}} \times \frac{187.77 \text{ g AgBr}}{1 \text{ mol AgBr}} = \frac{1.9\times10^{-9} \text{ g AgBr}}{1000 \text{ ml soln}}$$

10-83 In each case calculate K_c for each diagram as

$K_c = \frac{n-\text{butane}}{\text{isobutane}} = \frac{\bullet}{\circ}$. Only response (d) yields $K_c = 0.4$

10-85 Since $K_c = \frac{k_f}{k_r} = 1 \times 10^{-3}$, $k_r > k_f$.

10-87 (a) $[Sr^{2+}] = 5.8 \times 10^{-4}$ M, $[F^-] = 2 \times [Sr^{2+}] = 1.2 \times 10^{-3}$ M
$K_{sp} = [Sr^{2+}][F^-]^2 = 8.4 \times 10^{-10}$

(b) $$\left(\frac{0.100 \text{ mols } Sr(NO_3)_2}{1 \text{ L soln}}\right)\left(\frac{1 \text{ mol } Sr^{2+}}{1 \text{ mol } Sr(NO_3)_2}\right)(0.05000 \text{ L soln}) = 5.00 \times 10^{-3} \text{ mols } Sr^{2+}$$

$$[Sr^{2+}] = \frac{5.00 \times 10^{-3} \text{ mols } Sr^{2+}}{0.05000\text{L} + 0.05000\text{L}} = 0.0500 \text{ M } Sr^{2+}$$

$$\left(\frac{0.100 \text{ mols NaF}}{1 \text{ L soln}}\right)\left(\frac{1 \text{ mol } F^-}{1 \text{ mol NaF}}\right)(0.05000 \text{ L soln}) = 5.00 \times 10^{-3} \text{ mols } F^-$$

$$[F^-] = \frac{5.00 \times 10^{-3} \text{ mols } F^-}{0.1000\text{L}} = 0.0500 \text{ M } F^-$$

$Q = [Sr^{2+}][F^-]^2 = (0.0500)^3 = 1.25 \times 10^{-4} > K_{sp}$.
Since $Q > K_{sp}$, the reaction shifts to the left and SrF_2 will precipitate.

10-89 (a)

	$I_2(g)$	$\rightleftarrows$	$2\ I(g)$	$K_c = 3.8 \times 10^{-5}$ at 1000°C
initial	0.198 M		0	
change	$-\Delta c$		$2\Delta c$	
equil	$0.198-\Delta c$		$2\Delta c$	

$$\frac{(2\Delta c)^2(\Delta c)}{(0.400 - 2\Delta c)^2} = 3.8 \times 10^{-5}$$

Assume $\Delta c << 0.198$

$4\ \Delta c^2 = (0.198)(3.8 \times 10^{-5})$

$\Delta c^2 = 1.88 \times 10^{-6}$

$\Delta c = 1.37 \times 10^{-3}$

Check, $\frac{3.8 \times 10^{-4}}{0.400} \times 100\% = 0.69\%$, the assumption is valid.

$[I]_{eq} = 2\ \Delta c = 2.7 \times 10^{-3}$ $\quad [I_2]_{eq} = 0.198-\Delta c = 0.197$

Check, $\frac{(2\Delta c)^2(\Delta c)}{(0.400 - 2\Delta c)^2} = 3.8 \times 10^{-5}$, which agrees with the given value.

(b) ~0.197 At equilibrium almost all the iodine will be found in the form I_2. So having a starting concentration of 0.396 M of I, at equilibrium this will be converted mostly to I_2 which will then have a concentration of 0.198 M minus a small amount which is still in the monatomic form.

10-91

	$2\ NOCl(g)$	$\rightleftarrows$	$2\ NO(g)$	+	$Cl_2(g)$	$K_c = 1.6 \times 10^{-5}$ at 35°C
initial	1.0 M		0		0	
change	$-2\Delta c$		$2\Delta c$		Δc	
equil	$1.0-2\Delta c$		$2\Delta c$		Δc	

$$\frac{(2\Delta c)^2(\Delta c)}{(0.100-2\Delta c)^2} = 1.6 \times 10^{-5}$$

Assume $2\ \Delta c << 1.0$

$$\frac{4\Delta c^3}{(0.100)^2} = = 1.6 \times 10^{-5}$$

$$\Delta c^3 = \frac{3.4 \times 10^{-9}}{4} = 4.0 \times 10^{-6}$$

$\Delta c = 1.6 \times 10^{-2}$

Check, $\frac{2\left(9.5 \times 10^{-4}\right)}{0.100} \times 100\% = 3.2\%$, the assumption is valid.

$[NOCl]_{eq} = 0.97$ M $[NO]_{eq} = 0.032$ M $[Cl_2]_{eq} = 0.016$ M

10-93 (a) For each molecule of Ag_2SO_4 there will be two silver ions and one sulfate.

$[Ag^+]=1.5 \times 10^{-2} \times 2 = 3.0 \times 10^{-2}$ M

(b) $[SO_4^{2-}]=1.5 \times 10^{-2}$ moles

$K_{sp} = [Ag^+]^2 [SO_4^{2-}] = 1.4 \times 10^{-5}$

10-95 (a)

	$N_2F_4(g)$	$\rightleftarrows$	$2\ NF_2(g)$	$K_c=5.0 \times 10^{-9}$
initial	1.0 M		0	
change	Δc		$2\Delta c$	
equilibrium	$1.0 - \Delta c$		$2\Delta c$	

$$K_c = \frac{(2\Delta c)^2}{(1.0 - \Delta c)} = 5.0 \times 10^{-9}$$

Assume $\Delta c << 1.0$

$4\ \Delta c^2 = 5.0 \times 10^{-9}$

$\Delta c = 3.5 \times 10^{-5}$

Check, $\frac{3.5 \times 10^{-5}}{1.00} \times 100\% = 3.5 \times 10^{-3}\%$, the assumption is valid.

$[N_2F_4]_{eq} = 1.0 - 3.5 \times 10^{-5} = 1.0$ M; $[NF_2]_{eq} = 3.5 \times 10^{-5}$ M

(b) Since the reactants are favored over the products in this reaction, all of the product NF_2 would form N_2F_4 and the final concentrations of N_2F_4 would be 1.0 M.

10A-1 For the equation: $\dfrac{[0.125 - \Delta c]\,[2.40 - 2\Delta c]^2}{[0.200 + 2\Delta c]^2} = 1.3 \times 10^{-8}$

If the assumption is made that Δc is zero, one is left with a mathematical inequality. The change, Δc must be no larger than 0.125. This boundary condition is necessary to avoid having a negative concentration. One method of approximation is to set up a function $f(\Delta c) = 0$ and try various values for Δc until the solution of $f(\Delta c) = 0$ is attained. For example

$f(\Delta c) = 0 = (0.125 - \Delta c)\,(2.40 - \Delta c)^2 - 1.3 \times 10^{-8}\,(0.200 + 2\,\Delta c)^2$

Let $\Delta c = 0$, $f(\Delta c) = 0.719$

Let $\Delta c = 0.120$, $f(\Delta c) = 0.023$

Let $\Delta c = 0.124$, $f(\Delta c) = 0.0046$

Let $\Delta c = 0.1249$, $f(\Delta c) = 4.6 \times 10^{-4}$

Let $\Delta c = 0.124999$, $f(\Delta c) = 4.6 \times 10^{-6}$

Let $\Delta c = 0.124999999$, $f(\Delta c) = 1.99 \times 10^{-9}$

For most purposes $\Delta c = 0.1249$ would be considered a solution to the polynomial. It is recommended that this polynomial be solved using a computer algorithm.

10A-3 The assumption of a small change in concentration is doomed to failure when the initial concentrations are very different from the equilibrium concentrations. By the fact of being "very different" the changes must become "very much larger" to accomplish the adjustments required.

10A-5 (a) and (b) are unacceptable procedures since the value of K_c cannot be changed. It is a constant of the chemical system for a given temperature. The direction of the adjustment should make the value of Q_c become as near to that of K_c as possible. Response (c).

Chapter 11
Acids and Bases

11-1 An **acid** is a proton donor. A **base** is a proton acceptor.
Vinegar and lemon juice are examples of acids.
Lye is an example of a base.

11-3 Litmus will turn red in the presence of an acid. Litmus will remain blue in the presence of a base.

11-5 (a) and (b) are Arrhenius acids.

11-7 $HBr(g) + H_2O(l) \rightarrow H_3O^+(aq) + Br^-(aq)$
$H_2O(l) + NH_3(g) \rightarrow OH^-(aq) + NH_4^+(aq)$

11-9 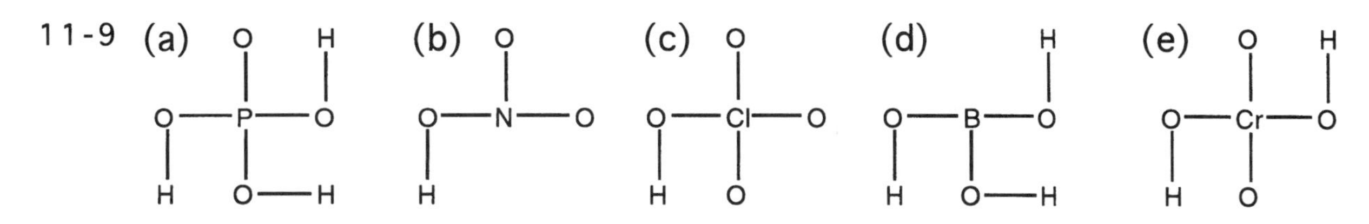

11-11 A **Brønsted base** is a substance that can accept a proton, i.e., a H^+ ion. The hydronium ion H_3O^+ with the positive charge will not add a second proton. H_3O^+ and BH_4^- are the only ions listed which cannot be Brønsted bases.

11-13 (a) $HSO_4^-(aq)$ + $H_2O(l)$ → $H_3O^+(aq)$ + $SO_4^{2-}(aq)$
Brønsted acid, Brønsted base → Brønsted acid, Brønsted base

(b) $CH_3CO_2H(aq)$ + $OH^-(aq)$ → $CH_3CO_2^-(aq)$ + $H_2O(l)$
Brønsted acid, Brønsted base → Brønsted base, Brønsted acid

(c) $CaF_2(s)$ + $H_2SO_4(aq)$ → $CaSO_4(aq)$ + 2 $HF(aq)$
Brønsted base, Brønsted acid → Brønsted base, Brønsted acid

(d) $HNO_3(aq)$ + $NH_3(aq)$ → $NH_4NO_3(aq)$
Brønsted acid, Brønsted base

The NH_4^+ ion of $NH_4NO_3(aq)$ is a Brønsted acid.
The NO_3^- ion of $NH_4NO_3(aq)$ is a Brønsted base.

(e) $LiCH_3(l)$ + $NH_3(l)$ → $CH_4(g)$ + $LiNH_2(s)$
CH_3^- from $LiCH_3(l)$ is a Brønsted base.
$NH_3(l)$ is a Brønsted acid in this reaction.
CH_4 is an acid and $LiNH_2$ is a base.

11-15

(a)	$HCl(aq)$ acid	+	$H_2O(l)$ base	→	$H_3O^+(aq)$ Acid	+	$Cl^-(aq)$ base	
(b)	$HCO_3^-(aq)$ base		$H_2O(l)$ acid	→	$OH^-(aq)$ Base	+	$H_2CO_3(aq)$ acid	
(c)	$NH_3(aq)$ base	+	$H_2O(l)$ acid	→	$OH^-(aq)$ Base	+	$NH_4^+(aq)$ acid	
(d)	$CaCO_3(s)$ base	+	2 $HCl(aq)$ acid	→	$Ca^{2+}(aq)$	+	$2Cl^-(aq)$ base	+ $H_2CO_3(aq)$ acid

11-17

(a)

formic acid formate ion

(b)

methanol methoxide

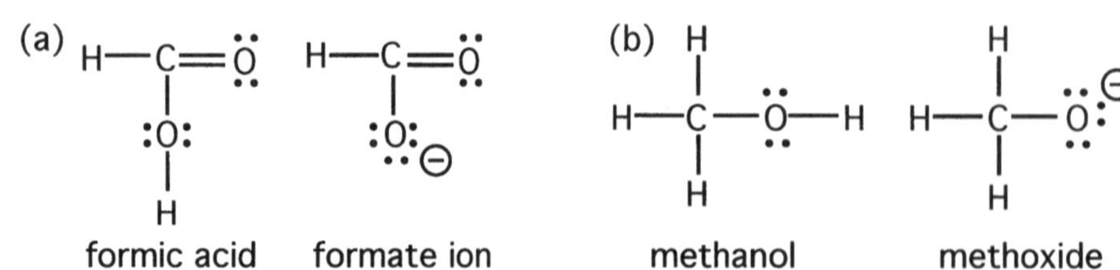

11-19 The conjugate base of a Brønsted acid will have a molecular formula differing from that of the acid by the loss of one H^+.

(a) H_2O (b) OH^- (c) O^{2-} (d) NH_3

11-21 The conjugate acid of a Brønsted base will have a molecular formula differing from that of the bases by addition of one H^+.

(a) OH^- (b) H_2O (c) H_3O^+ (d) NH_3

11-23 This statement is true because we know that water will dissociate in the following reaction: $H_2O\ (l) + H_2O\ (l) \rightarrow H_3O^+(aq) + OH^-(aq)$. Since one of the water molecules is donating a proton, this is a Brønsted acid. The other water molecule in this reaction is accepting the proton, and is therefore a Brønsted base .

11-25

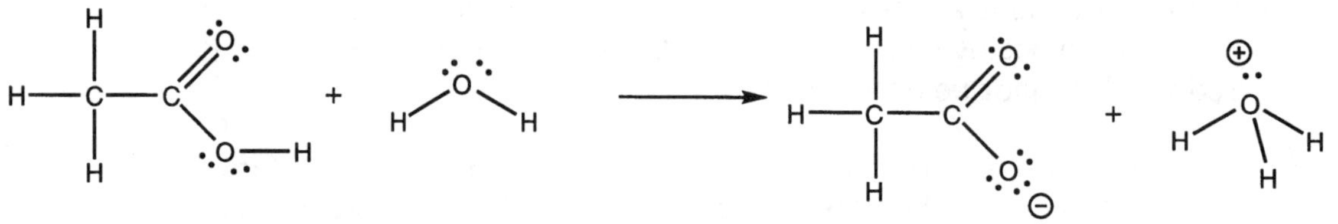

11-27 Water cannot be at equilibrium unless

$$K_w = [H_3O^+]\ [OH^-]$$

If water contained large quantities of both ions, the reaction product, K_w, would far exceed the limiting value of 1.0×10^{-14}.

11-29 Addition of a strong acid or base to water increases the concentration of either the H_3O^+ ion or the OH^- ion, respectively. This shifts the dissociation reaction of water, shown below, to the left. $2H_2O\ (l) \rightleftarrows H_3O^+\ (aq) + OH^-\ (aq)$

11-31 When an acid is added to water the two sources of H_3O^+ are from the reaction of the acid with water (the acid donates a proton) and from the dissociation of water. Similarly when a base is added to water it takes a proton from water molecules, producing OH^- ions. But these are also present from the dissociation of water.

11-33 $K_w = [H_3O^+]\ [OH^-] = 1.0 \times 10^{-14}$, therefore if $(1.0 \times 10^{-12})[OH^-] = 1.0 \times 10^{-14}$, then $[OH^-] = 0.010$ M.

11-35 $[H_3O^+] = 10^{-pH}$

pH = 3.72 then $[H_3O^+] = 1.9 \times 10^{-4}$ M

pH + pOH = 14.00 $[OH^-] = 10^{-pOH}$

pOH = 10.28 and $[OH^-] = 5.3 \times 10^{-11}$ M

11-37 When a strong acid is added to water, the hydronium ion (H_3O^+) concentration increases. The hydroxide ion (OH^-) concentration decreases. The pH of the solution also decreases.

11-39 The most acidic water solution will exhibit the smallest pH. The pH of 2.9 for the 0.10 M acetic acid solution indicates that it is the most acidic solution.

11-41 $pH=-\log([H_3O^+])$
$pH=-\log(1.5 \times 10^{-6}) = 5.8$
$pOH=14-pH=8.2$

11-43 $[H_3O^+] = 10^{-pH}$
pH = 3.8 then $[H_3O^+] = 1.6 \times 10^{-4}M$
pH + pOH = 14.00 $[OH^-] = 10^{-pOH}$
pOH = 10.2 and $[OH^-] = 6.3 \times 10^{-11}$ M

11-45 pH + pOH = 14.00, pH=11.1, pOH = 2.9
$[OH^-] = 10^{-pOH}$, $[OH^-] = 1.3 \times 10^{-3}$ M

11-47 Strong acids release protons nearly completely while weak acids release protons only to a limited degree. HCl is fully dissociated in water, while acetic acid is not.
Strong bases accept protons most readily while weak bases accept protons only to a limited degree. NaOH is fully dissociated in water, releasing all the hydroxide ions, whereas NH_3 will bind with water but not always release a proton.

11-49

	Acid	Ka
(a)	acetic acid	1.75×10^{-5}
(b)	boric acid	7.3×10^{-10}
(c)	chromic acid	9.6
(d)	formic acid	1.8×10^{-4}
(e)	hydrobromic acid	1×10^{9}

Both chromic acid and hydrobromic acid qualify as "strong acids" because their $K_a > 1$. The others are classified as weak acids.

11-51 For equal concentrations of weak acids in water solution, the larger the value of Ka the stronger the acid. Cl_2CHCO_2H is the strongest acid of those listed with a Ka = 5.1×10^{-2}. Response (d).

11-53 The stronger acid will generate a higher $[H_3O^+]$ concentration in water solution. The correct order of the listed compounds from the weakest to strongest acid is:

$HOAc < HNO_2 < HF < HOClO$

11-55 I^-, ClO_4^-, NO_3^- will not act as bases in water since they are the conjugates of strong acids, HI, $HClO_4$, and HNO_3.

11-57 According to the Brønsted model, the stronger acid will transfer a proton, becoming the weaker conjugate base. The stronger base will accept a proton, becoming a weaker conjugate acid. In the reaction of hydrogen chloride with water, the stronger proton donor, HCl, transfers the proton to the stronger base, water, creating the weaker conjugate acid, hydronium ion and the weaker conjugate base, chloride ion.

11-59 According to the Brønsted model, the stronger acid will transfer a proton, becoming the weaker conjugate base. The stronger base will accept a proton, becoming a weaker conjugate acid.

$$HBr(aq) + H_2O(l) \rightarrow H_3O^+(aq) + Br^-(aq)$$

The reaction above indicates that hydrogen bromide is a stronger acid than water, because it gives up its proton more readily.

11-61 $HCO_2^-(aq) + H_2O(l) \rightarrow HCO_2H(aq) + OH^-(aq)$

The fact that this reaction will not proceed as written indicates that water is not a stronger acid than formic acid and will not donate a proton to HCO_2^-. The reaction will proceed as:

$HCO_2H(aq) + OH^-(aq) \rightarrow H_2O(l) + HCO_2^-(aq)$

This indicates that OH^- is a stronger base than HCO_2^- and will accept a proton from formic acid.

11-63 Since strong acids and bases completely dissociate in water, their apparent strength is based solely upon their concentration in water rather than the value of the dissociation constant for the species. This is because the strength of the H_3O^+ or OH^- produced limits the acid or base strength. This limitation is called the **leveling effect.**

11-65 The H_3O^+ concentration in a strong acid solution depends on the concentration of the solution and not the value of K_a for the acid, because the acid can be assumed to be completely dissociated.

11-67 $AsO_4^{-3} < HAsO_4^{-2} < H_2AsO_4^- < H_3AsO_4$ As the charge gets less negative the acidity increases.

11-69 (d) H_2Te is the strongest acid because the H–X bond length is the longest, meaning that it is also the weakest bond and the easiest to dissociate in water.

11-71 $HOCH_3 < HOI < HOBr < HOCl$. The more electronegative the element attached to the oxygen the more likely that electron density will be pulled away from the hydrogen, making it more acidic.

11-73 $CH_3COOH < CH_2ClCOOH < CCl_3COOH$. The more chlorines attached the more likely that electron density will be pulled away from the hydrogen, making it more acidic.

11-75 $$\frac{0.568 \text{ g HCl}}{250 \text{ ml}} \times \frac{1 \text{ mol HCl}}{36.461 \text{ g HCl}} \times \frac{1000 \text{ ml}}{1 \text{ L}} = 0.0623 \frac{\text{mol HCl}}{\text{L}}$$

$pH = - \log [H_3O^+]$

$pH = 1.2$

$pOH = 12.8$

11-77 (a) $[H_3O^+]=1.0$, pH= 0

(b) $[H_3O^+]= \frac{0.568 \text{ g HCl}}{250 \text{ ml}} \times \frac{1 \text{ mol HCl}}{36.461 \text{ g HCl}} \times \frac{1000 \text{ ml}}{1 \text{ L}} = 0.0226$ M, pH=1.65

(c) $[H_3O^+]= \frac{0.568 \text{ g HCl}}{250 \text{ ml}} \times \frac{1 \text{ mol HCl}}{36.461 \text{ g HCl}} \times \frac{1000 \text{ ml}}{1 \text{ L}} = 0.0361$ M, pH=1.44

(d) $[H_3O^+]= \frac{0.568 \text{ g HCl}}{250 \text{ ml}} \times \frac{1 \text{ mol HCl}}{36.461 \text{ g HCl}} \times \frac{1000 \text{ ml}}{1 \text{ L}} = 0.0282$ M, pH=1.55

11-79 Assuming that HNO_3 is completely dissociated:
pH = - log $[H_3O^+]$ = - log (0.10) = 1.00
Using $K_a = 28$:

$$\frac{[H_3O^+]\,[NO_2^-]}{[HNO_3]} = \frac{[\Delta c]\,[\Delta c]}{[0.10 - \Delta c]} = 28$$

$\Delta c^2 = 28\,\Delta c - 2.8 = 0$, solving with the quadratic equation,

$$\Delta c = \frac{-28 + \sqrt{28^2 - 4(1)(-2.8)}}{2(1)} = 0.10 \frac{\text{mol } H_3O^+}{\text{L}}$$

pH = $-\log[H_3O^+]$ = -log(0.10) = 1.00

11-81 In solutions of strong acids or strong bases, the corresponding hydronium and hydroxide ion concentrations are completely determined. By definition, a strong acid or base fully dissociates in water solution. In solutions of very weak acids or bases, the dissociation of H_2O is a significant contribution to the hydronium and hydroxide ion concentrations.

11-83 For solutions of the same weak acid, the solution with the highest molarity also has the highest H_3O^+ ion concentration. $[H_3O^+] = \sqrt{K_aC_{HA}}$. The 0.10 M HOAc solution has the largest H_3O^+ ion concentration of the solutions listed. Response (a).

11-85 For the weak acid formic acid, $[H_3O^+] = (K_aC_{HA})^{1/2}$ and C_{HA} is the initial concentration of HA.

$$[H_3O^+] = \sqrt{(1.8 \times 10^{-4})(0.10)} = 4.2 \times 10^{-3} \text{ M} = [H_3O^+] = [HCO_2^-]$$

$$[HCO_2H] = 0.100 - 4.2 \times 10^{-3} = 9.6 \times 10^{-2} \text{ M}$$

11-87 For the weak acid phenolic acid, $[H_3O^+] = (K_aC_{HA})^{1/2}$ and C_{HA} is the initial concentration of HA.

$$[H_3O^+] = [PhO^-] = \sqrt{(1.0 \times 10^{-10})(0.0167)} = 1.3 \times 10^{-6} \text{ M}$$

11-89 Assume that the dissociation of the second proton on ascorbic acid does not contribute to the species in solution.

	$C_6H_6OH_2(aq)$	+ $H_2O(l)$ ⇄	$H_3O^+(aq)$	+	$C_6H_6OH^-(aq)$
Initial	0.10 M		0		0
Equil	0.10 M-2.8%(0.10M)		2.8%(0.10M)		2.8%(0.10M)

$$K_a = \frac{[C_6H_6OH^-][H_3O^+]}{[C_6H_6OH_2]} = \frac{[(0.028)(0.10)]^2}{(0.100 - 0.0028)} = 8.1 \times 10^{-5}$$

11-91 The dissociation of pure water has a pH of 7. Adding acid to it, even in small concentrations, will only bring the pH down. The error in this assumption is in forgetting the H_3O^+ present due to the dissociation of water.

11-93 pH= - log $[H_3O^+]$ = 4.27

$$K_a = \frac{[(0.071)(0.10)]^2}{(0.100 - 0.0071)} = 5.4 \times 10^{-4}$$

A 0.10 M solution of HNO_2 would have an $[H_3O^+]$ equal to the concentration of acid, and the pH would be 1.0

11-95 For a conjugate acid - base pair, $K_aK_b = K_w$.
Response (c).

11-97 Assume that the dissociation of the sodium formate is complete.

	$HCO_2^-(aq)$	+ $H_2O(l)$ ⇄	$HCO_2H(aq)$	+	$OH^-(aq)$
Initial	0.080 M		0		0
Equil	0.080 - Δc		Δc		Δc

$$K_b = \frac{K_w}{K_a} = \frac{1.0 \times 10^{-14}}{1.8 \times 10^{-4}} = 5.6 \times 10^{-11}$$

$$K_b = \frac{[HCO_2H][OH^-]}{[HCO_2^-]} = 5.6 \times 10^{-11}$$

$$\frac{\Delta c^2}{0.080 - \Delta c} = 5.6 \times 10^{-11}$$

Assume $\Delta c \ll 0.080$

$$\Delta c^2 = (0.080)\ 5.6 \times 10^{-11} = 4.5 \times 10^{-12}$$

$\Delta c = 2.1 \times 10^{-6}$ M. The assumption is valid.

$[OH^-]=[HCO_2H]=2.1 \times 10^{-6}M$

$[HCO_2^-]=0.080M$

11-99 Consider the dissociation of the weak acid HA.

	$OAc^-(aq)$	+	$H_2O(l)$	$\rightleftarrows$	$HOAc(aq)$	+	$OH^-(aq)$
Initial	0.756 M				0		0
Equilibrium	$0.756 - \Delta c$				Δc		Δc

$$K_b = \frac{K_w}{K_a} = \frac{1.0 \times 10^{-14}}{1.8 \times 10^{-5}} = 5.6 \times 10^{-10}$$

$$K_b = \frac{[HOAc][OH^-]}{[OAc^-]} = 5.6 \times 10^{-10}$$

$$\frac{\Delta c^2}{0.756 - \Delta c} = 5.6 \times 10^{-10}$$

Assume $\Delta c << 0.756$

$\Delta c^2 = 4.2 \times 10^{-10}$

$\Delta c = 2.0 \times 10^{-5}$ M. The assumption is valid.

$[OH^-] = 2.0 \times 10^{-5}$ M

pOH = 4.7

pH = 14.00 - 4.7 =9.3

11-101

	$CH_3NH_2(aq)$	+	$H_2O(l)$	$\rightleftarrows$	$CH_3NH_3^+(aq)$	+	$OH^-(aq)$
Equil	93.2%(0.10M)				6.8%(0.10M)		6.8%(0.10M)

$$K_b = \frac{[CH_3NH_3^+][OH^-]}{[CH_3NH_2]} = \frac{(0.0068)^2}{0.093} = 5.0 \times 10^{-4}$$

Methylamine is a stronger base than aqueous ammonia.

11-103 There are two ways to calculate the pH of a 0.10 M solution of methylamine given the information contained in problem 101.

1. The $[OH^-]$ is $\frac{6.8}{100}(0.10) = 0.0068$.

from which pOH is 2.17 and pH = 14.00-2.17 = 11.83.

2. K_b was calculated to be 5.0×10^{-4}.

$$K_b = \frac{[CH_3NH_3^+][OH^-]}{0.10 - \Delta c} = 5.0 \times 10^{-4}$$

$$\frac{\Delta c^2}{0.10 - \Delta c} = 5.0 \times 10^{-4}$$

Solving the quadratic equation, we get $\Delta c = [OH^-] = 0.0068$, for which we can calculate pH=11.8.

11-105

	$OAc^-(aq)$	+	$H_2O(l)$	$\rightleftarrows$	$HOAc(aq)$	+	$OH^-(aq)$
Initial	0.032 M				0		0
Equilibrium	0.032 - Δc				Δc		Δc

The concentration of AcO^- ion $= 0.016 \dfrac{\text{mol } Ca(OAc)_2}{\text{liter}} \times \dfrac{2 \text{ mol } OAc^-}{1 \text{ mol } Ca(OAc)_2} = 0.032$ M.

$$K_b = \frac{K_w}{K_a} = \frac{1.0 \times 10^{-14}}{1.8 \times 10^{-5}} = 5.6 \times 10^{-10}$$

$$K_b = \frac{[HOAc][OH^-]}{[OAc^-]} = 5.6 \times 10^{-10}$$

$$\frac{\Delta c^2}{0.032 - \Delta c} = 5.6 \times 10^{-10}$$

Assume $\Delta c << 0.032$

$\Delta c^2 = 1.8 \times 10^{-11}$

$\Delta c = 4.2 \times 10^{-6}$ M. The assumption is valid.

$[OH^-] = 4.2 \times 10^{-6}$ M.

pOH = 5.4

pH = 8.6

11-107 (a) NH_4Cl. NH_4^+] is a weak acid, whereas NO_3^- and F^- are conjugates of weak acids.

11-109 (a)NaH_2PO_4 (b)H_2CO_3 (c) $NaHSO_4$ (d) HNO_2

11-111 For the weak acid HF, $[H_3O^+] = (K_aC_{HA})^{1/2}$ where C_{HA} is the initial concentration of HA. $C_{HA} = 0.50$ M

$$[H_3O^+] = \sqrt{(1.8 \times 10^{-5})(0.10)} = 1.9 \times 10^{-2} \text{ M, pH} = 1.72$$

$$K_a = \frac{[HCO_2^-][H_3O^+]}{[HCO_2H]},\ [H_3O^+] = \frac{K_a[HCO_2H]}{[HCO_2^-]} = \frac{(1.8 \times 10^{-4})(0.10)}{1.0} = 7.2 \times 10^{-4},\ \text{pH} = 3.1$$

11-113 $$K_a = \frac{[NH_3][H_3O^+]}{[NH_4^+]},\ [H_3O^+] = \frac{K_a[NH_4^+]}{[NH_3]} = \frac{(5.6 \times 10^{-10})(0.10)}{0.1} = 5.6 \times 10^{-10},\ \text{pH} = 9.3$$

11-115 A **buffer** is composed of both the weak acid or base and its conjugate form at appreciable concentration. Addition of an acid consumes some of the conjugate base form buffering the influence of the addition. For addition of a base, some of the conjugate acid form of the buffering pair is consumed. The net result is that the hydronium ion concentration in the solution does not change very much when either an acid or a base is added to the solution.

11-117 To be an effective buffer, we seek systems which have a conjugate pair of substances related to each other by transfer of a proton. Neither should be exceptionally strong. Response (e).

11-119 Response (d) is a basic buffer.

11-121 (a), (c), and (d) will all go nearly to completion.

11-123 pH= 7.0 Equal moles of a strong acid and base have been added. They will neutralize each other.

11-125 (a) $HOBr(aq) + OH^-(aq) \rightleftarrows H_2O(l) + OBr^-(aq)$

(b) $HOAc(aq) + NH_3(aq) \rightleftarrows OAc^-(aq) + NH_4^+(aq)$

(c) $H_3O^+(aq) + OH^-(aq) \rightleftarrows 2\ H_2O(l)$

(d) $2\ HCOOH(aq) + Ba(OH)_2(s) \rightleftarrows 2\ HCOO^-(aq) + 2\ H_2O(l) + Ba^+(aq)$

Reactions (a), (b) and (c) will go to completion.

11-127 A **titration** is when a known quantity of one reagent is added to a second solution until the end point has been reached. The **end point** is when an indication is given that the equivalence point of the titration has been reached. The **equivalence point** in a titration is when the known amount of material added by titration is equal to the amount of material already present in the solution. That is, equivalent amounts of each substance have been added to the solution. The end point is determined by use of an **indicator,** which changes color when the end point has been reached.

11-129 The titration curve for the titration of the weak base NH_3 with strong acid would be as follows:

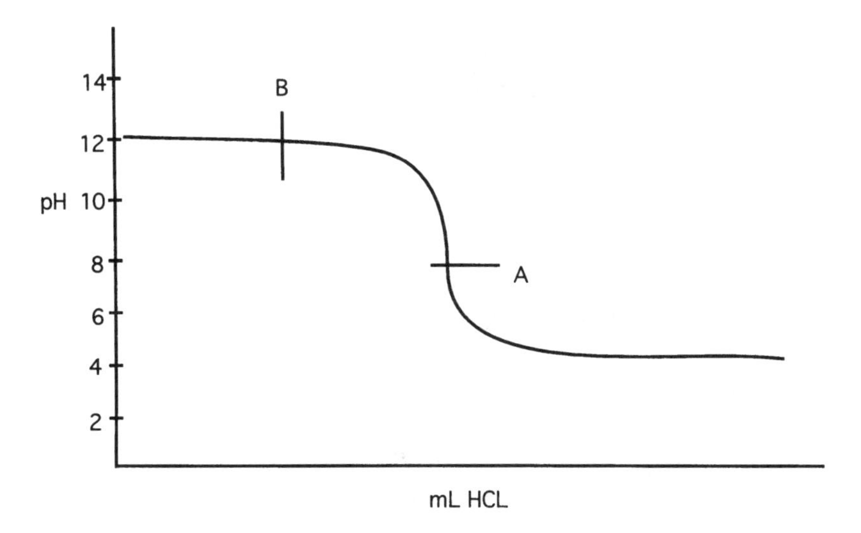

Point A is the equivalence point of the titration. Equal molar amounts of acid and base have reacted and the curve drops steeply.

Point B is the half - way point in the titration. At this point in the titration, the amounts of NH_3 and NH_4^+ are equal and the $[OH^-] = K_b$ for the base.

11-131 By definition:

$[H_3O^+] = 10^{-pH}$

The rain in Scotland had $[H_3O^+] = 10^{-2.4}$ mol/L = 0.004 M.

The acetic acid solution has $[H_3O^+] = 10^{-2.9}$ mol/L = 0.001M.

11-133 (a) False, each HCl that dissociates produces an H_3O^+ and a Cl^-.
(b) True
(c) True
(d) False, since HCl is a strong acid, it completely dissociates.

11-135 (a) The hydrogen attached to the more electronegative atom, oxygen, is acidic.
(b)

$$H{-}C({=}O){-}\ddot{O}{-}H + H{-}\ddot{O}{-}H \longrightarrow H{-}C({=}O){-}\ddot{O}:^{\ominus} + H{-}\overset{H}{\ddot{O}}{}^{\oplus}{-}H$$

(c)

	HCO_2H	+ $H_2O(l)$ ⇄	$H_3O^+(aq)$	+	$HCO_2^-(aq)$
Initial	0.10 M		0		0
Equil	0.10 - Δc		Δc		Δc

since pH = 2.37, $10^{-2.37} = [H_3O^+] = \Delta c = 4.3 \times 10^{-3}$

$$K_a = \frac{[HCO_2^-][H_3O^+]}{[HCO_2H]} = \frac{(4.3x10^{-3})(4.3x10^{-3})}{(0.10)} = 1.8 \times 10^{-4}$$

(d) The pH of the solution should be less than 2.37 because the addition of the Cl atom should cause an increase in the strength of the acid.

11-137 A list in order of strongest acid to strongest base will also be in order of increasing pH.
$HBr < HOBrO_2 < HOBrO < NaBr < NaBrO_3 < NaBrO_2$

11-139 (a) A dilute solution of a strong acid would consist of A^- and H_3O^+ in small amounts. Figure C matches this description.
(b) A concentrated solution of a weak acid would consist of a majority of HA and some H_3O^+ and A^-. Figure A matches this description.
(c) A buffer solution would consist of equal amounts of HA and A^- with some H_3O^+. The figure on the right matches this description.

11-141 Conjugate base: (c)Conjugate acid : (a)

11-143 (a) $Al_2O_3(s) + 6\ HCl(aq) \rightarrow 2\ AlCl_3(aq) + 3\ H_2O(l)$
(b) $CaO(s) + H_2SO_4(aq) \rightarrow CaSO_4(aq) + H_2O(l)$
(c) $Na_2O(s) + H_2O(l) \rightarrow 2Na^+(aq) + 2HO^-(aq)$
(d) $MgCO_3(s) + 2\ HCl(aq) \rightarrow H_2CO_3(aq) + MgCl_2(aq)$
(e) $NaOH(s) + H_3PO_4(aq) \rightarrow 3\ H_2O(aq) + Na_3PO_4(aq)$

11-145 A. a) $HBr(aq) + H_2O(l) \rightarrow H_3O^+(aq) + Br^-(aq)$

b) $[H_3O^+] = 1.0$ M

c) pH = 0

d) $[OH^-] = 1.0 \times 10^{-14}$

e) $Br^-(aq)$

B. a) $NH_3(aq) + H_2O(l) \rightarrow NH_4^+(aq) + OH^-(aq)$

b) $K_b = \dfrac{[NH_4^+][OH^-]}{[NH_3]} = 1.8 \times 10^{-5}$,

$K_b = \dfrac{[NH_4^+][OH^-]}{[NH_3]} = 1.8 \times 10^{-5}$

c) $[NH_3] = 1.0 - 4.2 \times 10^{-3} = 1.0$ M

d) $[NH_4^+] = 4.2 \times 10^{-3}$ M

e) pH = 14 - pOH = 11.6

C. a) $[Br^-] = 1.0$ M

b) $Br^-(aq) + H_2O(l) \rightarrow HBr(aq) + OH^-(aq)$

The equilibrium constant will be low, because Br^- is a conjugate base of a weak acid.

c) pH = 7.0

11-147 (a) $HNO_2(aq) + H_2O(l) \rightarrow NO_2^-(aq) + H_3O^+(aq)$

(b) $K_a = \dfrac{[HCO_2^-][H_3O^+]}{[HCO_2H]} = \dfrac{(4.3 \times 10^{-3})(4.3 \times 10^{-3})}{(0.10)} = 5.1 \times 10^{-4}$

$\Delta c = 2.3 \times 10^{-2}$ M.

pH = 1.65

(c) $K_a = \dfrac{[HCO_2^-][H_3O^+]}{[HCO_2H]} = \dfrac{(4.3 \times 10^{-3})(4.3 \times 10^{-3})}{(0.10)}$

$\Delta c = 4.4 \times 10^{-6}$ M

pOH=5.35, pH=8.65

d) pH=0, strong acid

e) A, because the nitric acid will completely dissociate.

11-149 a) pH = 1.0, strong acid

b) pH = 7.0, conjugate base of strong acid

c) 1 < pH < 7, this is a weak acid

d) 7 < pH < 14, this is a weak base (conjugate of the weak acid)

e) pH = 14, strong base

11-151 Increasing acidity

KOH	strong base
$KClO_3$	conjugate base of weak acid
KI	conjugate base of strong acid
NH_4ClO_4	weak acid and conjugate base of strong acid
$HClO_3$	weak acid
$HClO_4$	strong acid

11-153 At the start of the titration, $K_a = \frac{[OAc^-][H_3O^+]}{[HOAc]} = \frac{\Delta c \cdot \Delta c}{.100} = 1.8 \times 10^{-5}$

$\Delta c = 1.3 \times 10^{-3}$ M.

pH = 2.87

After adding 20.0 mL of 0.10 M NaOH, there will be 0.003 moles of HOAc left in 70.0 mls, the [HOAc]= 0.0429 M. There will also be 0.002 moles of OAc^- in the 70.0 mls, the $[OAc^-]$= 0.0286 M. $K_a = \frac{[OAc^-][H_3O^+]}{[HOAc]} = \frac{0.0286 \cdot [H_3O^+]}{0.0429} = 1.8 \times 10^{-5}$

$[H_3O^+] = 2.7 \times 10^{-5}$ M, pH=4.57.

At the equivalence point when 50 mL of 0.10 M NaOH has been added, there will be 0.0050 moles of OAc^- in 100.0 mls of solution, $[OAc^-]$= 0.050 M

$$K_b = \frac{K_w}{K_a} = 5.56 \times 10^{-10} = \frac{[HOAc][OH^-]}{[OAc^-]} = \frac{\Delta c \cdot \Delta c}{0.050}$$

$\Delta c = 5.3 \times 10^{-6}$ M

pOH=5.28, pH=8.72

After adding 60.0 ml of the base the base concentration remaining is 0.001 moles of OH^- in 110 ml, $[OH^-]$=0.0091 M. This will also be a greater contribution to the pH than the weak base, OAc^- , which is also in the solution. pH= 12.0.

11-155 (b) HSO_4^- is the conjugate base of the strong acid sulfuric acid. H_3O^+ is a strong acid. The remainder of the ions are conjugate bases of weak acids.

11A-1 $K_{a1} = \frac{[HCO_3^-][H_3O^+]}{[H_2CO_3]} = 4.5 \times 10^{-7}$ $K_{a2} = \frac{[CO_3^{2-}][H_3O^+]}{[HCO_3^-]} = 4.7 \times 10^{-11}$

Assume $[H_2CO_3]$ at equilibrium = 0.100 M.

Assume $[HCO_3^-]=[H_3O^+] = \Delta c$

$\frac{\Delta c^2}{0.100} = 4.5 \times 10^{-7}$, $\Delta c^2 = 4.5 \times 10^{-8}$, $\Delta c = 2.1 \times 10^{-4}$M

$[HCO_3^-]=[H_3O^+] = 2.1 \times 10^{-4}$ M

$$\frac{[CO_3^{2-}][H_3O^+]}{[HCO_3^-]} = 4.7 \times 10^{-11}$$

$$\frac{[CO_3^{2-}]\left(2.1 \times 10^{-4}\right)}{\left(2.1 \times 10^{-4}\right)} = 4.7 \times 10^{-11} \text{ M}$$

$[CO_3^{2-}] = 4.7 \times 10^{-11}$ M

11A-3 Check the assumption that $[H_3O^+] \cong [HM^-]$

$[H_3O^+]$ and $[HM^-]$ are $\frac{1.87 \times 10^{-3}}{0.250} \times 100 = 0.75\%$ of the initial value of H_2M.

Therefore the assumption that $[H_2M] = H_2M$ initial is valid.

The concentration of $M^{2-} = 2.1 \times 10^{-8}$ M. This is about 89,000 times smaller than $[HM^-]$ and $[H_3O^+]$. So the assumption that nearly all of the HM^- and H_3O^+ come from the first dissociation is valid.

11A-5 (a) 9.3×10^{-2} M < 1.9 M no

(b) 9.3×10^{-2} M < 1.9 M yes

(c) 9.3×10^{-2} M $= 9.3 \times 10^{-2}$ M no

(d) 9.3×10^{-2} M $= 9.3 \times 10^{-2}$ M yes

(e) 9.3×10^{-2} M $= 9.3 \times 10^{-2}$ M no

Responses (b) and (d)

11A-7 Assume stepwise dissociation of oxalic acid,

$$K_{a1} = \frac{[H_3O^+][HC_2O_4^-]}{[H_2C_2O_4]} = 5.4 \times 10^{-2}$$

$$K_{a2} = \frac{[H_3O^+][C_2O_4^{2-}]}{[HC_2O_4^-]} = 5.4 \times 10^{-5}$$

Working with the first dissociation,

$$\frac{\Delta c^2}{1.25 - \Delta c} = 5.4 \times 10^{-2}$$

First approximation: $\frac{(\Delta c')^2}{1.25} = 5.4 \times 10^{-2}$ $\Delta c' = 0.260$

Second approximation: $\frac{(\Delta c'')^2}{1.25 - 0.260} = 5.4 \times 10^{-2}$ $\Delta c'' = 0.231$

Third approximation: $\frac{(\Delta c''')^2}{1.25 - 0.260} = 4.5 \times 10^{-3}$ $\Delta c''' = 0.235$

Check, $\frac{(0.235)^2}{(1.25 - 0.235)} = 5.44 \times 10^{-2}$ M, which agrees with the value given.

$\Delta c = 0.24$ M $= [H_3O^+] = [HC_2O_4^-]$, and $[H_2C_2O_4] = 1.01$ M

11A-9

$$C_2O_4^{2-}(aq) + H_2O(l) \rightleftarrows HC_2O_4^-(aq) + OH^-(aq)$$

$$K_{b1} = \frac{K_w}{K_{a2}} = \frac{1.0 \times 10^{-14}}{5.4 \times 10^{-5}} = 1.9 \times 10^{-10}$$

$$HC_2O_4^-(aq) + H_2O(l) \rightleftarrows H_2C_2O_4^-(aq) + OH^-(aq)$$

$$K_{b2} = \frac{K_w}{K_{a1}} = \frac{1.0 \times 10^{-14}}{5.4 \times 10^{-2}} = 1.9 \times 10^{-13}$$

Working with K_{b1},

Assume $[OH^-] \cong [HC_2O_4^-] = \Delta c$ and $[C_2O_4^{2-}] \cong C_{Na_2C_2O_4}$

$$\frac{\Delta c^2}{0.028 - \Delta c} = 1.9 \times 10^{-10}$$

$\Delta c^2 = (0.028)(1.9 \times 10^{-10}) = 5.3 \times 10^{-12}$

$\Delta c = 2.3 \times 10^{-6} M = [OH^-] = [HC_2O_4^-]$

$[C_2O_4^{2-}] = 0.028$ M

pOH = 5.6

pH = 8.4

11A-11 Assume stepwise dissociation of phosphoric acid,

$$K_{a1} = \frac{[H_3O^+][HC_2O_4^-]}{[H_2C_2O_4]} = 5.4 \times 10^{-2}$$

$$K_{a2} = \frac{[H_3O^+][C_2O_4^{2-}]}{[HC_2O_4^-]} = 5.4 \times 10^{-5}$$

$$K_{a2} = \frac{[H_3O^+][C_2O_4^{2-}]}{[HC_2O_4^-]} = 5.4 \times 10^{-5}$$

Working with the first dissociation,

$$\frac{\Delta c^2}{1.25 - \Delta c} = 5.4 \times 10^{-2}$$

First approximation: $\frac{(\Delta c')^2}{0.20} = 7.1 \times 10^{-3}$ $\Delta c' = 0.0377$

Second approximation: $\frac{(\Delta c'')^2}{0.20 - 0.0377} = 7.1 \times 10^{-3}$ $\Delta c'' = 0.0339$

Third approximation: $\frac{(\Delta c''')^2}{0.20 - 0.0339} = 7.1 \times 10^{-3}$ $\Delta c''' = 0.0343$

Check, $\frac{(0.235)^2}{(1.25 - 0.235)} = 5.44 \times 10^{-2}$ M, which agrees with the value given.

$\Delta c = 0.0343$ M $= [H_3O^+] = [H_2PO_4^-]$, and $[H_3PO_4] = 0.166$ M

Now looking at the second dissociation and substituting values for $[H_3O^+]$ and $[H_2PO_4^-]$:

$$K_{a2} = \frac{[H_3O^+][C_2O_4^{\ 2-}]}{[HC_2O_4^{\ -}]} = 5.4 \times 10^{-5}$$, so $[HPO_4^{2-}]=6.3\times10^{-8}$. Given the difference between concentrations of HPO_4^{2-} and $H_2PO_4^-$ our assumption that the further dissociation to form H_3O^+ is so small that the $[H_3O^+]$ is relatively unchanged from the first dissociation.

We take this assumption one step further in the third dissociation:

$$K_{a2} = \frac{[H_3O^+][C_2O_4^{\ 2-}]}{[HC_2O_4^{\ -}]} = 5.4 \times 10^{-5}, \ [PO_4^{3-}]=7.7\times10^{-19}$$

11A-13 When solid $NaHCO_3$ is put into water, it ionizes to give Na^+ and HCO_3^- ions. HCO_3^- ion is a stronger base than it is an acid. The HCO_3^- ions will accept a proton. The only substance in solution that can donate a proton to HCO_3^- is the water molecule. This leaves an excess of OH^- ions in solution. This is why the solution is basic.

11A-15 0.10 M H_3PO_4 $K_{a1} = 7.1 \times 10^{-3}$

H_3PO_4 is acidic $[H_3O^+]= \Delta c$

$\Delta c^2 = 7.1 \times 10^{-4}$

$\Delta c = 2.7 \times 10^{-2} = [H_3O^+]$ pH = 1.6

0.10 M $NaHPO_4$

The $H_2PO_4^-$ ion is acidic $K_{a2} = 6.3 \times 10^{-8}$

$[H_3O^+] = \Delta c$

$\Delta c^2 = 6.3 \times 10^{-9}$ and $\Delta c = 7.9 \times 10^{-5}$ pH = 4.1

0.10 M Na_2HPO_4

The HPO_4^{2-} ion is basic $K_{b2} = 1.6 \times 10^{-7}$

$[OH^-] = \Delta c$

$\Delta c^2 = 1.6 \times 10^{-8}$ and $\Delta c = 1.3 \times 10^{-4}$

pOH = 3.9 and pH = 10.1

0.10 M Na_3PO_4

The PO_4^{3-} ion is basic $K_{b1} = \frac{K_w}{K_{a3}} = \frac{1.0 \times 10^{-14}}{4.2 \times 10^{-13}} = 0.024$ $[OH^-] = \Delta c$

$\Delta c^2 = 2.4 \times 10^{-3}$ and $\Delta c = 4.9 \times 10^{-2}$

pOH = 1.3 and pH = 12.7

11A-17 $K_{b2} = \frac{K_w}{K_{a1}} = \frac{1 \times 10^{-14}}{1.7 \times 10^{-2}} = 5.9 \times 10^{-13}$

$K_{a2} > K_{b2}$, the solution of $NaHSO_3$ will be acidic.

Chapter 12
Oxidation-Reduction Reactions

12-1 **Oxidation** occurs when an atom or molecule loses electrons in the course of a reaction. **Reduction** occurs when an atom or molecule gains electrons in the course of a reaction.

12-3 (a) reduction
(b) oxidation
(c) reduction
(d) oxidation

12-5 (a)CrO_4^{2-}, +6 (b) $Cr_2O_7^{2}$, +6 (c) CrO_2, +4

12-7 (c) LiH

12-9 (a)

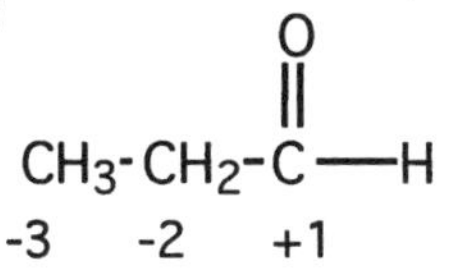

O
||
CH_3-CH_2-C—H
-3 -2 +1

(b)

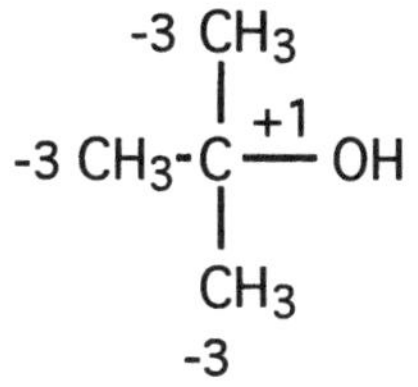

-3 CH_3
+1
-3 CH_3-C— OH
CH_3
-3

(c)

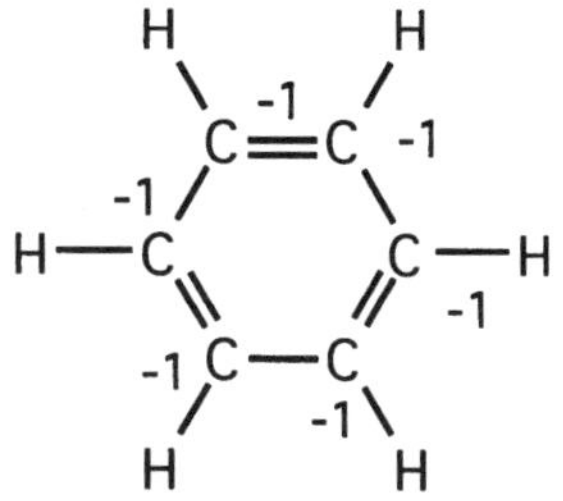

H H
-1
C=C -1
-1
H—C C—H
-1
-1 C—C
-1
H H

12-11

(a)

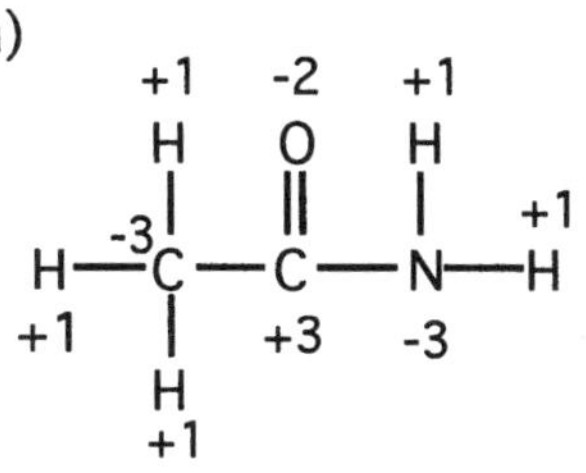

+1 -2 +1
H O H
-3 || +1
H—C—C—N—H
+1 +3 -3
H
+1

(b)

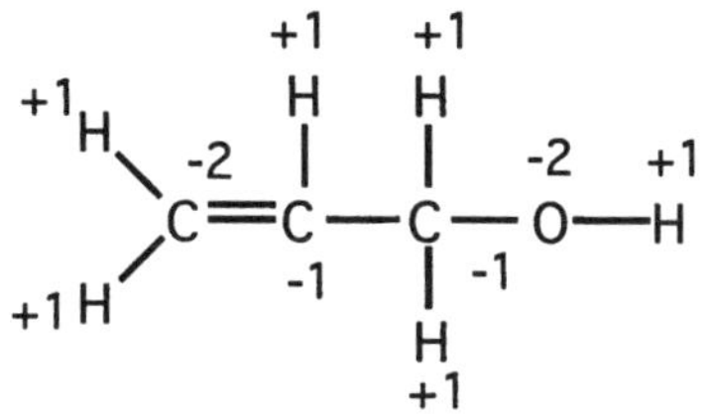

+1 +1
+1 H H H
-2 -2 +1
C=C—C—O—H
-1 -1
+1 H H
+1

(c)

+1 +1 +1
H H O -2 H
-3 -2 || -2 +1
H—C—C—C—O—C—H
+1 +3 -2
H H H +1
+1 +1

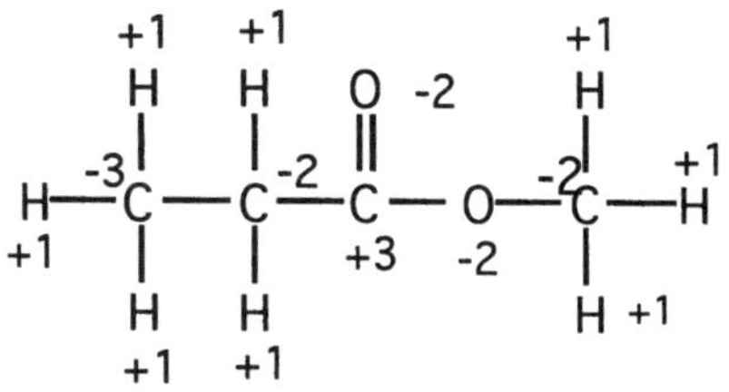

12-13 (b)

12-15 (a) and (b) are oxidation reactions. In (a), CH3OH is oxidized, and (b) CH3CH2CHO is oxidized.

12-17 This is not a redox reaction.

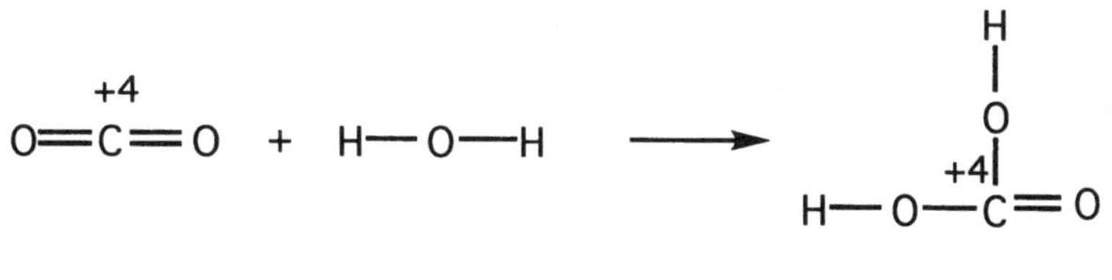

+4
O=C=O + H—O—H ⟶

H
|
O
+4|
H—O—C=O

12-19 In the reaction below the nitrogen is being reduced by the addition of three electrons.

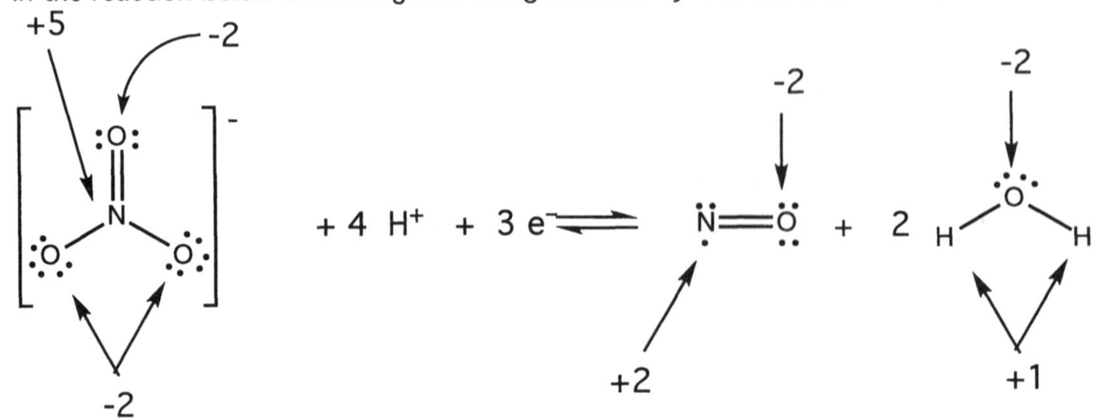

12-21 $CH_3CH_2OH \rightarrow CH_3CHO + 2e^- + 2H^+$
The net oxidation state of the carbon next to the oxygen atom in ethanol is assigned a value of -1. The carbon atom next to the oxygen atom in acetaldehyde is assigned a net oxidation state of + 1. The oxidation state of carbon increases from - 1 to + 1. This is equivalent to the loss of two electrons, or the removal of two hydrogen atoms. Ethanol is oxidized to acetaldehyde.

12-23

succinate: $^{\ominus}O(O=)C^{+3}-C_aH_2^{-2}-C_aH_2^{-2}-C^{+3}(=O)O^{\ominus}$

fumarate: $^{\ominus}O(O=)C^{+3}-C_aH^{-1}=C_aH^{-1}-C^{+3}(=O)O^{\ominus}$

The two C_as change from a -2 oxidation state in succinate to a -1 state in fumarate; therefore, this is a redox reaction.

12-25 The iron is oxidized from the 0 oxidation state to the +2 oxidation state. Oxygen is reduced from the 0 oxidation state to the -2 oxidation state. This is a redox reaction.

12-27 The **cathode** is the electrode of the electrochemical cell where reduction occurs. The **anode** is the electrode where oxidation occurs. A **cation** is a positive ion. An **anion** is a negative ion.

12-29 When oxidation occurs at the anode, a substance will give up electrons at the electrode surface. When reduction occurs at the cathode, the substance to be reduced accepts electrons at the electrode surface.

12-31 (a) $Al(s) + Cr^{3+}(aq) \rightleftarrows Al^{3+}(aq) + Cr(s)$
(b) $2\ Fe^{2+}(aq) + I_2(aq) \rightleftarrows 2\ Fe^{3+}(aq) + 2\ I^-(aq)$
(c) $Cr(s) + 3\ Fe^{3+}(aq) \rightleftarrows Cr^{3+}(aq) + 3\ Fe^{2+}(aq)$
(d) $Zn(s) + 2\ H^+(aq) \rightleftarrows Zn^{2+}(aq) + H_2(g)$

12-33 (a) The Al is oxidized and Cr^{3+} is reduced. So the Al is the reducing agent while the Cr^{3+} is the oxidizing agent.
(b) One of the Cr^{2+} is reduced to Cr and the two others are oxidized to Cr^{3+}. In this case the Cr^{2+} is both the oxidizing and reducing agent.
(c) The Fe is oxidized and Cr^{3+} is reduced, so the Fe is the reducing agent and Cr^{3+} is the oxidizing agent.
(d) The H_2 is being oxidized and $Cr^{3+}$is reduced, so the H_2 is the reducing agent and Cr^{3+} is the oxidizing agent.

12-35 Yes, because when it oxidizes another substance it will gain electrons, in the +4 oxidation state carbon will not lose any more electrons.

12-37 An oxidizing agent must itself be reduced in the process. Chloride ion (a) and zinc metal (d) cannot be further reduced. The calcium ion of calcium hydride (e) does not become reduced except under very select conditions. Only bromine (b) and iron(3+) (c) are expected to be oxidizing agents.

12-39 All of the substances listed can be either oxidizing or reducing agents under the proper conditions, although (c) H_2O is the least likely to be a reducing agent.

12-41 When an atom is in the lowest, or most negative, oxidation state it cannot take in any more electrons and will therefore not be an oxidizing agent. So it can only act as a reducing agent.

12-43 When using a table of relative reactivity such as Table 12.1, there is a preferred direction for combination of spontaneous reactions. In this situation, spontaneous reactions occur in a counter clockwise manner. A reaction higher in the table proceeds from right to left when coupled with a reaction farther down in the table (which must proceed from left to right).
(a) The reverse reaction should occur. So the reaction will not occur as written.
(b) The reverse reaction should occur. So the reaction will not occur as written.
(c) This reaction should occur as written.

12-45 (a) $Cr^{3+}(aq)$, $Fe^{3+}(aq)$ (or $Fe^{2+}(aq)$)
(b) $Mg^{2+}(aq)$, $Cr^{2+}(aq)$
(c) These ions will not react.

12-47 (a) These ions can coexist and will not react.
(b) These ions can coexist and will not react.
(c) Chromium will be oxidized to + 3 while triiodide will be reduced to iodide.

12-49 The better oxidizing agents are found at the bottom of Table 12.1. Fluorine reacting in acidic solution, response (d), is the strongest oxidizing agent of the group.

12-51 Set up a galvanic salt bridge cell with a copper metal electrode in a 1M Cu^{2+} solution and an iron metal electrode in a 1M Fe^{3+} solution. Allow the reaction to occur. Observe the electrodes to see at which electrode additional metal is deposited. The electrode at which metal is being deposited is being reduced, thus the other solution is a better reducing agent. By examining the reduction potentials we can determine that the copper electrode will undergo the reduction reaction, depositing copper metal on the electrode surface. The iron metal will be the reducing agent. Therefore, iron metal is a better reducing agent than copper.

12-53 The strongest oxidizing agent will be the substance with the most positive standard reduction potential. Response (e).

12-55 The better reducing agent will have the most positive standard reduction potential.
(a) K
(b) Sn
(c) Ag^+

12-57 Standard conditions for electrochemical measurements have been defined as 1 M concentration for all solutions and a partial pressure of 1 atm for all gases. Although standard-state measurements can be made at any temperature, they are often taken at 25°C.

12-59 The magnitude of E° for an oxidation-reduction reaction indicates the tendency for a given reaction to proceed in the direction described. The potential developed by the cell reflects the relative strengths of a pair of oxidizing agents and the relative strengths of a pair of reducing agents. Only the potential "difference" can be measured.

12-61 The sign of the standard state potential for an oxidation-reduction reaction must change when the direction in which the reaction is written is reversed. The magnitude remains the same.

12-63 Oxidation-reduction reactions should occur as written (are spontaneous) when the corresponding cell potential is positive. Reactions (a), (b) and (c) should occur as written. Reaction (d) should occur in the reverse direction.

12-65 $E^\circ_{overall} = E^\circ_{ox} + E^\circ_{red}$
(a) $E^\circ_{overall}$ = + 0.7628 V + 0.0000 V = + 0.7628 V
(b) $E^\circ_{overall}$ = + 0.74 V + 0.770 V = + 1.51 V
(c) $E^\circ_{overall}$ = 1.18 V + (-2.375) = -1.20 V
Reactions (a) and (b) should occur as written.
Reaction (c) should not occur as written.

12-67 In the cell the cathode is made of copper and the anode is made of zinc. The reaction at the anode is the oxidation of Zn, $Zn\ (s) \rightarrow Zn^{2+}\ (aq) + 2\ e^-$. The reaction at the cathode is the reduction of copper (II) ion, $Cu^{2+}\ (aq) + 2\ e^- \rightarrow Cu\ (s)$. The standard cell potential is $E^\circ_{cell} = E^\circ_{ox} + E^\circ_{red}$ = 0.76 V + 0.34 V = 1.10 V. During the reaction, copper will be deposited at the cathode and zinc will be dissolved from the anode into the solution.

12-69 $E^\circ_{overall} = E^\circ_{ox} + E^\circ_{red}$ = 0.036 V - 1.706 V = -1.67 V
The reaction should not occur spontaneously as written.

12-71 $E^\circ_{overall} = E^\circ_{ox} + E^\circ_{red}$ = -0.158 V + 0.522 V = 0.364 V
The reaction should occur spontaneously as written.

12-73 The reduction half-reaction for all three reactions will be $2\ H^+ + 2\ e^- \rightarrow H_2$. The oxidation half-reaction that will give the largest cell potential for the overall redox reaction will be the most reactive. Mg will be the most reactive.

12-75 Copper, gold, mercury, platinum and silver can be found in their metallic state in nature. This observation is in harmony with the relatively large positive standard reduction potentials associated with the metals. Lithium metal, on the other hand, has a very large negative reduction potential. For this reason it is more likely to be found in its oxidized state rather than the metallic state.

12-77 The cell potential is a direct measure of the driving force for the electron transfer reaction. As the system approaches equilibrium, the driving force associated with the reacting system becomes smaller. When the system reaches equilibrium, there is no driving force on or by the system. The cell potential is zero.

12-79 Response (a), the cell potential equals zero volts describes an oxidation-reduction reaction at equilibrium.

12-81 $E^\circ_{cell} = E^\circ_{ox} + E^\circ_{red} = -0.770\ V + 0.88\ V = 0.11\ V$

at pH = 10, pOH = 4, and $[OH^-] = 10^{-4}$ M

$$E_{cell} = E^\circ_{cell} - \frac{0.02569}{n} \ln Q_c$$

$$E_{cell} = 0.11\ V - \frac{0.02569}{2} \ln \frac{[1.0]^2[10^{-4}]^2}{[1.0][1.0]^2} = 0.35\ V$$

12-83 $E^\circ_{cell} = E^\circ_{ox} + E^\circ_{red} = -\ 0.682\ V + 1.491\ V = 0.809\ V$

$$E_{cell} = 0.809 - \frac{0.02569}{n} \ln \frac{[O_2]^5\ [Mn^{2+}]^2}{[MnO_4^-]^2[H_2O_2]^5\ [H^+]^6}$$

As the pH increases, $[H^+]$ decreases and the value of Q_c will increase, E_{cell} will decrease. If $[H_2O_2]$ increases, the Q_c will decrease and E_{cell} will increase.

12-85 $E^\circ_{cell} = E^\circ_{ox} + E^\circ_{red} = 0.7628\ V + (-0.409\ V) = 0.354\ V$

at T=298K, $K = e^{\frac{nFE^\circ}{RT}}$, where n=2,

$K = e^{27.6} = 9.4 \times 10^{11}$

12-87 A voltaic cell is one in which a spontaneous chemical reaction drives electric current through a circuit. An electrolytic cell is one through which an electric current is passed to produce a chemical reaction.

12-89 An electrolytic cell could not "run" a motor since electricity must be used to "run" an electrolytic cell. A voltaic cell can be used to drive a motor, or similarly a voltaic cell can drive an electrolytic cell.

12-91 $1.0000\ C \times \frac{1\ mol\ Ag}{1\ mol\ e^-} \times \frac{107.87\ g\ Ag}{1\ mol\ Ag} = 1.1180 \times 10^{-3}\ g\ Ag$

12-93 $1.0 \text{ mol Na} \times \frac{1 \text{ mol } e^-}{1 \text{ mol Na}} \times \frac{96485 \text{ C}}{1 \text{ mol } e^-} \times \frac{\text{amp s}}{1 \text{ C}} \times \frac{1}{5.0 \text{ amp}} \times \frac{1 \text{ h}}{3600 \text{ s}} = 5.4 \text{ h}$

$1.0 \text{ mol Mg} \times \frac{2 \text{ mol } e^-}{1 \text{ mol Mg}} \times \frac{96485 \text{ C}}{1 \text{ mol } e^-} \times \frac{\text{amp s}}{1 \text{ C}} \times \frac{1}{5.0 \text{ amp}} \times \frac{1 \text{ h}}{3600 \text{ s}} = 11 \text{ h}$

$1.0 \text{ mol Al} \times \frac{3 \text{ mol } e^-}{1 \text{ mol Al}} \times \frac{96485 \text{ C}}{1 \text{ mol } e^-} \times \frac{\text{amp s}}{1 \text{ C}} \times \frac{1}{5.0 \text{ amp}} \times \frac{1 \text{ h}}{3600 \text{ s}} = 16 \text{ h}$

12-95 $1.000 \times 10^6 \text{ g } Cl_2 \times \frac{1 \text{ mol } Cl_2}{70.906 \text{ g } Cl_2} \times \frac{2 \text{ mol } e^-}{1 \text{ mol } Cl_2} \times \frac{96485 \text{ C}}{1 \text{ mol } e^-} = 2.722 \times 10^9 \text{ C}$

12-97 $2.56 \text{ h} \times \frac{3600 \text{ s}}{1 \text{ h}} \times \frac{1 \text{ C}}{\text{amp s}} \times 1.34 \text{ amp} \times \frac{1 \text{ mol } e^-}{96485 \text{ C}} \times \frac{1 \text{ mol } H_2}{2 \text{ mol } e^-} \times \frac{2.0158 \text{ g } H_2}{1 \text{ mol } H_2}$
$= 0.129 \text{ g } H_2$

$2.56 \text{ h} \times \frac{3600 \text{ s}}{1 \text{ h}} \times \frac{1 \text{ C}}{\text{amp s}} \times 1.34 \text{ amp} \times \frac{1 \text{ mol } e^-}{96485 \text{ C}} \times \frac{1 \text{ mol } O_2}{4 \text{ mol } e^-} \times \frac{31.9988 \text{ g } O_2}{1 \text{ mol } O_2}$
$= 1.023 \text{ g } O_2$

$\frac{\text{mass } O_2}{\text{mass } H_2} = \frac{1.023 \text{ g}}{0.129 \text{ g}} = \frac{8}{1}$

12-99 $Al^{3+} + 3 e^- \rightarrow Al$

$2Br^- \rightarrow Br_2 + 2e^-$

$2AlBr_3 \rightarrow 2Al + 3Br_2$

coulombs used $= 10.5 \text{ h} \times \frac{3600 \text{ s}}{1 \text{ h}} \times \frac{1 \text{ C}}{\text{amp s}} \times 20.0 \text{ amp} = 7.56 \times 10^5 \text{ C}$

grams of Al electroplated $= 10.5 \text{ h} \times \frac{3600 \text{ s}}{1 \text{ h}} \times \frac{1 \text{ C}}{\text{amp-s}} \times 20.0 \text{amp} \times \frac{1 \text{ mol } e^-}{96485 \text{ C}} \times \frac{1 \text{ mol Al}}{3 \text{ mol Al}} \times$
$\frac{26.982 \text{ g Al}}{1 \text{ mol Al}} = 70.5 \text{ g Al}$

Likewise,

$7.56 \times 10^5 \text{ C} \times \frac{1 \text{ mol } e^-}{96485 \text{ C}} \times \frac{1 \text{ mol } Br_2}{2 \text{ mol } e^-} \times \frac{159.808 \text{ g } Br_2}{1 \text{ mol } Br_2} = 626 \text{ g } Br_2$

$\frac{\text{mass } Br_2}{\text{mass Al}} = \frac{626 \text{ g}}{70.5 \text{ g}} = 8.88$

12-101 The number of coulombs involved is

$$2.0 \text{ h} \times \frac{3600 \text{ s}}{1 \text{ h}} \times \frac{1 \text{ C}}{\text{amp s}} \times 10.0 \text{ amp} = 7.20 \times 10^4 \text{ C}$$

This corresponds to $7.20 \times 10^4 \text{ C} \times \frac{1 \text{ mol e}^-}{96485 \text{ C}} = 0.746 \text{ mol e}^-$

(a) $K^+ + e^- \rightarrow K$

$$0.746 \text{ mol e}^- \times \frac{1 \text{ mol K}}{1 \text{ mol e}^-} \times \frac{39.098 \text{ g K}}{1 \text{ mol K}} = 29.2 \text{ g K}$$

(b) $Ca^{2+} + 2 e^- \rightarrow Ca$

$$0.746 \text{ mol e}^- \times \frac{1 \text{ mol Ca}}{2 \text{ mol e}^-} \times \frac{40.078 \text{ g Ca}}{1 \text{ mol Ca}} = 14.9 \text{ g Ca}$$

(c) $Sc^{3+} + 3 e^- \rightarrow Sc$

$$0.746 \text{ mol e}^- \times \frac{1 \text{ mol Sc}}{3 \text{ mol e}^-} \times \frac{44.956 \text{ g Sc}}{1 \text{ mol Sc}} = 11.2 \text{ g Sc}$$

Response (a)

12-103 The number of mols of electrons is

$$16.5 \text{ h} \times \frac{3600 \text{ s}}{1 \text{ h}} \times \frac{1 \text{ C}}{\text{amp s}} \times 1.00 \text{ amp} \times \frac{1 \text{ mol e}^-}{96485 \text{ C}} = 0.616 \text{ mol e}^-$$

The mol Ce = $21.6 \text{ g Ce} \times \frac{1 \text{ mol Ce}}{140.12 \text{ g Ce}} = 0.154 \text{ mol Ce}$

Ratio $\frac{\text{mol Ce}}{\text{mol e}^-} = \frac{0.154}{0.616} = 0.25$

$\frac{\text{mol e}^-}{\text{mol Ce}} = \frac{0.616}{0.154} = 4$

Therefore, for every mol Ce there are 4 mol e^- and the cerium ion has an oxidation state of 4^+. The formula is $CeCl_4$, x = 1, y = 4.

12-105 $15 \text{ min} \times \frac{60 \text{ s}}{1 \text{ min}} \times \frac{1\text{C}}{1 \text{ amp s}} \times 2.50 \text{ amp} \times \frac{1 \text{ mol e}^-}{96485 \text{ C}} = 0.0233 \text{ mol e}^-$

mol Au = $1.53 \text{ g Au} \times \frac{1 \text{ mol Au}}{196.97 \text{ g Au}} = 7.77 \times 10^{-3} \text{ mol Au}$

$\frac{\text{mol e}^-}{\text{mol Au}} = \frac{0.0233}{0.00777} = 2.99 \Rightarrow 3$

The oxidation number of gold is (3+).

12-107 (a)-(c)

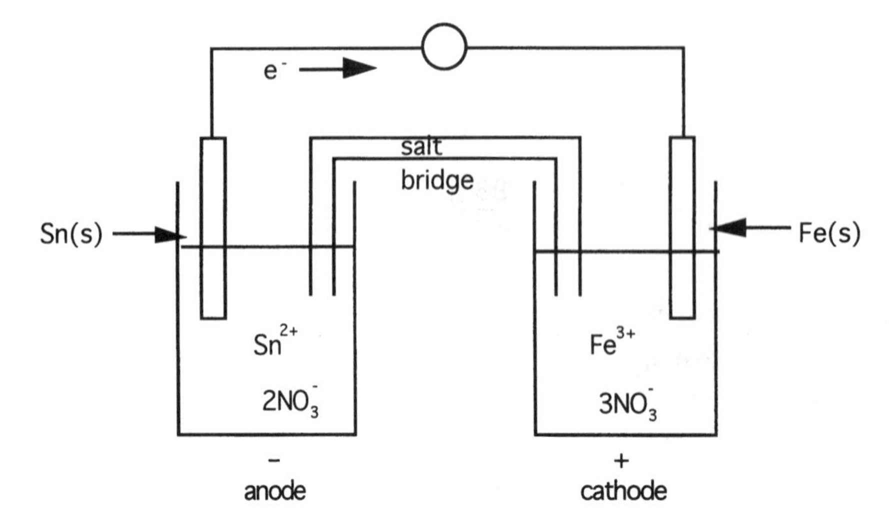

(d) $3Sn(s) + 2Fe^{3+}(aq) \rightarrow 3Sn^{2+}(aq) + 2Fe(s)$
(e) $E_{cell}^{\circ} = +0.10V$

12-109 (a) Yes: iron is reduced (Fe^{+3} to Fe^{0}) and carbon is oxidized (C^{+2} to C^{+4}).
(b) Yes: phosphorous is reduced (P^{5+} to P^{0}) and carbon is oxidized(C^{0} to C^{2+}).
(c) No: this is a precipitation reaction.

12-111 (a) This is a redox reaction. The C is oxidized from -4 to +4 and the O is reduced from 0 to -2.
(b) This is a redox reaction. The C is oxidized from +2 to +4 and the Pb is reduced from +2 to 0.
(c) There is no redox reaction.
(d) This is a redox reaction. The C's in ethylene are reduced from -2 to -3 and the H is oxidized from 0 to +1.

12-113 Only (b) is true

12-115 (a) Ag
(b) $E_{cell}^{\circ} = 1.56$ V
(c) The Zn cell is the anode and the Ag cell is the cathode.
(d) Electrons flow from Zn to Ag.
(e) $2\ Ag^{+}(aq) + Zn(s) \rightarrow 2\ Ag(s) + Zn^{2+}(aq)$
(f) At T=298K, $K = e^{\frac{nFE^{\circ}}{RT}}$, where n=2,
$K = e^{121.5} = 5.8 \times 10^{52}$

12-117 $3H_2C_2O_4(aq) + 2CrO_4^{2-}(aq) + 10H^{+}(aq) \rightarrow 6\ CO_2(g) + 2Cr^{3+}(aq) + 8H_2O(l)$

$$40.0\ \text{mL}\ CrO_4^{2-} \times \frac{0.0250\ \text{mol}\ CrO_4^{2-}}{1000\ \text{mL}} \times \frac{3\ \text{mol}\ H_2C_2O_4}{2\ \text{mol}\ CrO_4^{2-}} \times \frac{1}{0.0100\ \text{L}} = 0.150\ \text{mol}\ H_2C_2O_4$$

12-119 $5Fe^{2+}(aq) + MnO_4^-(aq) + 8H^+(aq) \rightarrow Mn^{2+}(aq) + 4 H_2O (l) + 5Fe^{3+}(aq)$

$$3.2 \text{ mL } MnO_4^- \times \frac{3.0 \text{ mol } MnO_4^-}{1000 \text{ mL}} \times \frac{5 \text{ mol } Fe^{2+}}{1 \text{ mol } MnO_4^-} \times \frac{1 \text{ mol } FeCl_2}{1 \text{ mol } Fe^{2+}} \times \frac{126.753 \text{ g } FeCl_2}{1 \text{ mol } FeCl_2}$$

$= 6.1$ g $FeCl_2$

12A-1 $3P_4(s) + 10KClO_3(s) \rightarrow 3P_4O_{10}(s) + 10KCl(s)$

The second reaction is not as easy to balance by inspection. The balanced equation is:

$C_2H_5OH(aq) + 2 Cr_2O_7^{2-} (aq) + 16 H^+(aq) \rightarrow 2 CO_2(g) + 4 Cr^{3+}(aq) + 11 H_2O(l)$

12A-3 (a) $5 Fe^{2+}(aq) + MnO_4^-(aq) + 8 H^+(aq) \rightarrow 5Fe^{3+}(aq) + Mn^{+2}(aq) + 4 H_2O(l)$

(b) $5 S_2O_3^{2-}(aq) + 8 MnO_4^-(aq) + 14 H^+(aq) \rightarrow 10 SO_4^{2-}(aq) + 8 Mn^{+2}(aq) + 7 H_2O(l)$

(c) $5 PbO_2(s) + 4 H^+(aq) + 2 Mn^{2+}(aq) \rightarrow 5 Pb^{2+}(aq) + 2 H_2O(l) + 2 MnO_4^-(aq)$

(d) $2 MnO_4^-(aq) + 6 H^+(aq) + 5 SO_2(g) + 2 H_2O(l) \rightarrow 2 Mn^{2+}(aq) + 5 H_2SO_4(aq)$

12A-5 (a)

For the oxidation half-reaction: $4(Cr(s) \rightarrow Cr^{3+}(aq) + 3e^-)$

For the reduction half-reaction: $3(4e^- + 4H^+(aq) + O_2(g) \rightarrow 2H_2O(l))$

Overall reaction: $4Cr(s) + 12H^+(aq) + 3O_2(g) \rightarrow 4Cr^{3+}(aq) + 6H_2O(l)$

(b)

For the oxidation half reaction: $6(Fe^{2+}(aq) \rightarrow Fe^{3+}(aq) + e^-)$

For the reduction half-reaction: $6e^- + 14H^+(aq) + Cr_2O_7^{2-}(aq) \rightarrow 2Cr^{3+}(aq) + 7H_2O(l)$

Overall reaction: $14H^+(aq) + Cr_2O_7^{2-}(aq) + 6Fe^{2+} \rightarrow 2Cr^{3+}(aq) + 6Fe^{3+}(aq) + 7H_2O(l)$

(c)

For the oxidation half-reaction: $10(4H_2O(l) + Cr^{3+}(aq) \rightarrow HCrO_4^-(aq) + 7H^+(aq) + 3e^-)$

For the reduction half-reaction: $3(10e^- + 12H^+(aq) + 2BrO_3^-(aq) \rightarrow Br_2(aq) + 6H_2O(l))$

Overall reaction: $22H_2O(l) + 10Cr^{3+}(aq) + 6BrO_3^-(aq) \rightarrow$

$10HCrO_4^-(aq) + 34H^+(aq) + 3Br_2(aq)$

12A-7 (a)HIO_3 (aq) + HI(aq) $\rightarrow$ I_2(aq)
oxidation half-reaction: HI(aq) $\rightarrow$ I_2(aq)
reduction half-reaction: HIO_3 (aq) $\rightarrow$ I_2(aq)

For the oxidation half-reaction: 5(2HI(aq) $\rightarrow$ I_2(aq) + $2H^+$ (aq) + $2e^-$)

For the reduction half-reaction: $10e^-$ + $10H^+$(aq) + $2HIO_3$ (aq) $\rightarrow$ I_2(aq) + 6 H_2O(l)

Overall reaction: HIO_3 (aq) + 5HI(aq) $\rightarrow$ $3I_2$(aq) + $3H_2O$(l)

(b)SO_2 (g) + H_2S(g) $\rightarrow$ S_8(s) + H_2O(l)

oxidation half-reaction: H_2S(g) $\rightarrow$ S_8(s)
reduction half-reaction: SO_2 (g) $\rightarrow$ S_8(s)

For the oxidation half-reaction: 2($8H_2S$(g) $\rightarrow$ S_8(s) + $16H^+$ (aq) + $16e^-$)

For the reduction half-reaction: $32e^-$ + $32H^+$(aq) + $8SO_2$ (g) $\rightarrow$ S_8(s) + 16 H_2O(l)

Overall reaction: $8SO_2$ (g) + $16H_2S$(g) $\rightarrow$ $3S_8$(s) + $16H_2O$(l)

12A-9 (a)3Cu(s) + $2HNO_3$(aq) + $6H^+$(aq) $\rightarrow$ $3Cu^{2+}$(aq) + 2NO(g) + $4H_2O$(l)

(b)Cu(s) + $2HNO_3$(aq) + $2H^+$(aq) $\rightarrow$ Cu^{2+}(aq) + $2NO_2$(g) + $2H_2O$(l)

12A-11 (a)NO(g) + MnO_4^-(aq) $\rightarrow$ NO_3^-(aq) + MnO_2(s)
(b)
For the oxidation half-reaction: $6OH^-$(aq) + $2NH_3$(aq) $\rightarrow$ N_2(g) +$6H_2O$(l) + $6e^-$

For the reduction half-reaction: 2($3e^-$ + $2H_2O$(l) +MnO_4^-(aq) $\rightarrow$ MnO_2(s)+$4OH^-$(aq))

Overall reaction: $2NH_3$(g) + $2MnO_4^-$(aq) $\rightarrow$ N_2(g) +$2H_2O$(l) +$2MnO_2$(s) +$2OH^-$(aq)

12A-13 (a)2Cr(s) + $2OH^-$(aq) + $6H_2O$(l) $\rightarrow$ $2Cr(OH)_4^-$(aq) + $3H_2$(g)

(b)$4OH^-$(aq) + $2Cr(OH)_3$(s) + $3H_2O_2$(aq) $\rightarrow$ $2CrO_4^{2-}$(aq) + $8H_2O$(l)

12A-15 4Au(s) + $2H_2O$(l) + $8CN^-$(aq) + O_2(g) $\rightarrow$ $4Au(CN)_2^-$ (aq) + $4OH^-$(aq)

12A-17 (a)

For the oxidation half-reaction: $4OH^-(aq) + Cl_2(g) \rightarrow 2OCl^-(aq) + 2H_2O(l) + 2e^-$

For the reduction half-reaction: $2e^- + Cl_2(g) \rightarrow 2Cl^-(aq)$

Overall reaction: $2OH^-(aq) + Cl_2(g) \rightarrow OCl^-(aq) + H_2O(l) + Cl^-(aq)$

(b)

For the oxidation half-reaction: $12OH^-(aq) + Cl_2(g) \rightarrow 2ClO_3^-(aq) + 6H_2O(l) + 10e^-$

For the reduction half-reaction: $5(2e^- + Cl_2(g) \rightarrow 2Cl^-(aq))$

Overall reaction: $6OH^-(aq) + 3Cl_2(g) \rightarrow ClO_3^-(aq) + 3H_2O(l) + 5Cl^-(aq)$

(c)

For the oxidation half-reaction: $3(8OH^-(aq) + P_4(s) \rightarrow 4H_2PO_2^-(aq) + 4e^-)$

For the reduction half-reaction: $12e^- + 12H_2O(l) + P_4(s) \rightarrow 4PH_3(g) + 12OH^-(aq)$

Overall reaction: $12OH^-(aq) + 4P_4(s) + 12H_2O(l) \rightarrow 12H_2PO_2^-(aq) + 4PH_3(g)$

(d)

For the oxidation half-reaction: $2OH^-(aq) + H_2O_2(aq) \rightarrow O_2(g) + 2H_2O(l) + 2e^-$

For the reduction half-reaction: $2e^- + H_2O(l) + H_2O_2(aq) \rightarrow H_2O(l) + 2OH^-(aq)$

Overall reaction: $2H_2O_2(aq) \rightarrow O_2(g) + 2H_2O(l)$

12A-19 (a) The carbon on CH_4 is oxidized and the O_2 is reduced.

(b) The double bonded carbons on propene are reduced and the H_2 is oxidized.

12A-21 When a lead acid battery is discharged PbO_2 is reduced to $PbSO_4$. When it is charged up again the reverse reaction is driven through electrolysis.

12A-23 The dry cell battery, as its name implies, has a lot less water present in the electrolyte solution. The ability to make the cell a lot smaller makes this kind of battery a lot more useful for smaller applications and devices.

12A-25 A fuel cell is an electrochemical cell in which a controlled "burn" or oxidation of a fuel with oxygen produces an electrical voltage in the same way a battery does.

12A-27 $2Fe(s) + O_2(g) \rightarrow 2FeO(s)$

$4Fe(s) + 3O_2(g) \rightarrow 2Fe_2O_3(s)$

Of these two the cell potential for the second reaction is greater.

12A-29 The $[H^+]$ is much less than 1.0 M therefore the reduction potential based upon the Nernst equation will be much less than the standard reduction potential.

12A-31 Rust can be better described as a hydrated oxide, $Fe_2O_3 \cdot 3H_2O$. It is also a mixture of oxides rather than just Fe_2O_3.

12A-33 Galvanized iron corrodes much more slowly than iron by itself. The more active zinc becomes the anode relative to the iron metal. This protects the iron metal from being oxidized. This process is called cathodic protection.

12A-35 At the anode the following reaction will take place: $2F^{-}(aq) \rightarrow F_2(g) + 2e^{-}$

At the cathode the following reaction will take place: $Ca^{2+}(aq) + 2e^{-} \rightarrow Ca(s)$

12A-37 Water has a more positive standard reduction potential than Na^{+} ions. Therefore, water will be reduced at the cathode, forming $H_2(g)$ and OH^{-} ions. At the anode, Cl^{-} ion is oxidized to Cl_2 gas. There are various reasons why Cl^{-} ion is preferentially oxidized to Cl_2 rather than H_2O to $O_2(g)$. The most compelling of these is the very high over-potential for the oxidation of water.

12A-39 When aqueous magnesium chloride is electrolyzed, magnesium(2+) ions are reduced at the cathode and chloride(1-) ions are oxidized at the anode. Response (c).

12A-41 If nickel sulfate were used as the salt in an electrolysis cell, the "over voltage" associated with production of hydrogen would allow some of the nickel(2+) ions to be reduced. Depending upon the relative acidity of the electrolyte and the concentration of the nickel ions, the amount of hydrogen released could be controlled.

12A-43 The high negative standard reduction potential associated with potassium(1+) ions suggests that the cathode reaction for electrolysis of a sodium bromide solution in water will be hydrogen. Since chlorine gas was produced when electrolysis of a chloride containing solution was undertaken, we should expect to obtain bromine at the anode of the electrolysis cell using an aqueous solution of potassium bromide as the electrolyte. Response (c).

12A-45 Aluminum metal cannot be prepared by the electrolysis of an aqueous solution of the aluminum(3+) ion. The reduction potential required is so great that only hydrogen gas will be liberated.

Chapter 13
Chemical Thermodynamics

Note to student: Unless specifically directed by the problem, all thermodynamic values used in the solutions for this chapter are for the atom combination rather than formation.

13-1 In referring to chemical reactions **spontaneous** refers to the fact that the reaction is favorable. Barring kinetic considerations, it will proceed from reactants to products.

13-3 Yes, the freezing of liquid water is spontaneous below 0°C but not spontaneous above 0°C.

13-5 Gaseous water has more entropy, because the gas has less constraints on its motion and more disorder in the system.

13-7 As warm water molecules come in contact with the penny and transfer energy to the penny. As this happens the average kinetic energy (temperature) of the penny increases and the average kinetic energy of the water decreases until they are equal. When the system has a uniform temperature the disorder as measured by the distribution of energies (see chapter 6) is at its greatest.

13-9 If ΔS is negative, the system becomes more ordered. Response (d). A gaseous reactant forms a solid product.

13-11 Entropy increases as the disorder of the system increases. Response (c).

13-13 An entropically favored process is one in which ΔS_{sys} is positive. The process is not spontaneous if the ΔS_{univ} is negative as a result of the process.

13-15 All solutions have a concentration of 1.0 M.
All gases have a partial pressure of 1.0 bar (0.9869 atm).
At temperature of 25°C.

13-17 If the crystal is in perfect order and there is no inherent disorder (entropy).

13-19 Gas has the highest entropy, because there is the least amount of structure and order in that state of matter. Solids generally have a definite structure, and liquids, while lacking definite shape, do not have as great a freedom of motion as gases.

13-21 If $\Delta S°$ is positive it means that entropy increased from reactants to products. So the products must have more entropy than the reactants. The opposite is true when $\Delta S°$ is negative.

13-23 $\Delta S° = 2034.6 + 6 \times 320.57 - 4 \times 1374 = -1538$ J/mol_{rxn}K
Entropy is most likely not a driving force in the reaction.

13-25 $\Delta S° = 3 \times 116.972 - 2 \times 244.24 = -137.56$ J/mol_{rxn}K
The negative $\Delta S°$ indicates that the products of the reaction are more ordered than the reactants.

13-27 The standard-state Gibbs free energy change ($\Delta G°$) refers to a system under a unique set of standard conditions. ΔG refers to a system observed under any other of a wide variety of conditions.

13-29 $\Delta G° = \Delta H° - T\Delta S°$ for a reaction involving standard state species.
A negative value of $\Delta G°$ is associated with a spontaneous reaction. A spontaneous reaction will always result when $\Delta H° < 0$ and $\Delta S° > 0$.
Response (b).

13-31 $NH_3\,(g) \rightleftarrows NH_3\,(aq)$

$\Delta S° < 0$ because the gaseous phase is less ordered than the solution phase. $\Delta H° < 0$ because the increased hydrogen bonding potential should enhance the interactions in the water solution, but should not disrupt the water structure to any great extent. As the solution is heated, the ammonia gas is expelled. This happens because more $NH_3(aq)$ has enough energy to break the hydrogen bonds and become $NH_3(g)$.

13-33 $Fe_2O_3\,(s) + 2Al\,(s) \rightarrow Al_2O_3\,(s) + 2Fe\,(s)$

$\Delta H° = 2404.3 + 2 \times 326.4 - 3076.0 - 2 \times 416.3 = -851.5\ kJ/mol_{rxn}$
$\Delta S° = 756.75 + 2 \times 136.21 - 761.33 - 2 \times 153.21 = -38.58\ J/mol_{rxn}K$
The entropy change in this reaction is unfavorable, however the enthalpy change is so large that it is the driving force behind the reaction.

13-35 $2KMnO_4(s) + 5H_2O_2(aq) + 6H^+(aq) \rightarrow 2K^+(aq) + 2Mn^{2+}(aq) + 5O_2(g) + 8H_2O(l)$

$\Delta H°_{rxn} = 2 \times 2203.8 + 5 \times 1124.81 + 6 \times 217.65$
$-2 \times 341.62 - 2 \times 501.5 - 5 \times 498.340 - 8 \times 970.30 = -602.8\ kJ/mol_{rxn}$
$\Delta S°_{rxn} = 2 \times 806.50 + 5 \times 407.6 + 6 \times 114.713$
$-2 \times 57.8 - 2 \times 247.3 - 5 \times 116.972 - 8 \times 320.57 = 579.7\ J/mol_{rxn}K$

13-37 $PCl_5(g) + 4H_2O(l) \rightarrow H_3PO_4(aq) + 5HCl(g)$

$\Delta H° = 1297.9 + 4 \times 970.30 - 3241.7 - 5 \times 431.64 = -220.8\ kJ/mol_{rxn}$
$\Delta S° = 624.60 + 4 \times 320.57 - 1374 - 5 \times 93.003 = 68\ J/mol_{rxn}K$
All three, enthalpy change, entropy change, and Le Chatelier's principle, are driving forces in this reaction. $\Delta H°$ is negative, $\Delta S°$ is positive, and removal of HCl (g) will drive the reaction in the direction of product formation.

13-39 $3Fe(s) + 4H_2O(l) \rightarrow Fe_3O_4(s) + 4H_2(g)$

$\Delta G° = \Delta H° - T\Delta S°$
$\Delta H° = 3 \times 416.3 + 4 \times 970.30 - 3364.0 - 4 \times 435.30 = 24.9\ kJ/mol_{rxn}$
$\Delta S° = 3 \times 153.21 + 4 \times 320.57 - 1039.3 - 4 \times 98.742 = 307.6\ J/mol_{rxn}K$
$\Delta G° = 24.9\ kJ/mol_{rxn} - T(0.3076\ kJ/mol_{rxn}K)$
At the higher temperatures of "white hot metal," the $T\Delta S°$ term in the expression for free energy change will dominate, giving a negative free energy change. The reaction will be spontaneous at the high temperature.

13-41 In 1 M acid, the reaction of Zn is $Zn(s) + 2H^+(aq) \rightleftarrows Zn^{2+}(aq) + H_2(g)$

ΔG° = 95.145 + 2 x 203.247 - 242.21 - 406.494 = -147.06 kJ/mol_{rxn}
ΔH° = 130.729 + 2 x 217.65 - 284.62 -435.30 = -153.89 kJ/mol_{rxn}
ΔS° = -23.1 $J/mol_{rxn}K$
In water,
$Zn(s) + 2H_2O(l) \rightarrow Zn^{2+}(aq) + 2OH^-(aq) + H_2(g)$
ΔG° = 95.145 + 2 x 875.354 - 242.21 - 2 x 592.222 - 406.494
= 12.71 kJ/mol_{rxn}
This reaction is not spontaneous.
ΔH° = 130.729 + 2 x 970.30 - 284.62 - 2 x 696.81 - 435.30 = -42.2 kJ/mol_{rxn}
ΔS° = -184.4 $J/mol_{rxn}K$

The products of this reaction are less stable than the reactants as shown by ΔG°. Both reactions have favorable enthalpy changes but the reaction with water has such an unfavorable entropy change that the reaction does not proceed.

13-43 $SiH_4(s) \rightarrow Si(s) + 2H_2(g)$
ΔH° = 1291.9 - 455.6 - 2 x 435.30 = -34.3 kJ/mol_{rxn}
ΔS° = 422.20 - 149.14 - 2 x 98.742 = 75.58 $J/mol_{rxn}K$
ΔG° = -34.3 kJ/mol_{rxn} - T(0.07558 $kJ/mol_{rxn}K$)
The reaction is spontaneous. Since $\Delta H < 0$ increasing the temperature would make the equilibrium constant smaller, shifting the reaction to the left.

13-45

	ΔH°(kJ/mol_{rxn}) at 298 K	ΔS°($J/mol_{rxn}K$) at 298 K	ΔG°(kJ/mol_{rxn})
(a)	-565.97	-173.00	-514.42
(b)	483.64	90.01	+456.79
(c)	-164.1	148.66	-208.4
(d)	23.39	-12.5	+27.1

(a) ΔG° will become more positive as T increases since $\Delta S^\circ < 0$.
(b) ΔG° will become more negative as T increases since $\Delta S^\circ > 0$.
(c) ΔG° will become more negative as T increases since $\Delta S^\circ > 0$.
(d) ΔG° will become more positive as T increases since $\Delta S^\circ < 0$.

13-47 ΔG° will increase linearly with increasing temperature.

13-49 For $H_2O(s) \rightarrow H_2O(l)$
at -10°C $\Delta G > 0$
0°C $\Delta G = 0$
+10°C $\Delta G < 0$

13-51 The entropy increase associated with transition of water to the vapor phase is positive. The enthalpy change is endothermic.

13-53 $\Delta G° = \Delta H° - T\Delta S°$

$\Delta G°_{500} = -484\ kJ/mol_{rxn} - 500\ K(-88\ J/mol_{rxn}K) = -440\ kJ/mol_{rxn}$

$\Delta G°_{500} = -484\ kJ/mol_{rxn} - 1000\ K(-88\ J/mol_{rxn}K) = -396\ kJ/mol_{rxn}$

The calculated $\Delta G°$ values are more negative than the correct values. This indicates that either the $\Delta H°$ has a smaller magnitude at the higher temperatures or the $\Delta S°$ has a larger magnitude at the higher temperatures.

13-55 (a) $\Delta G° = -1882.25 - (-1040.156) - 2 \times (-406.494) = -29.11\ kJ/mol_{rxn}$

(b) $\Delta G° = 2 \times (-532.0) - 3 \times (-463.462) = 326.4\ kJ/mol_{rxn}$

(c) $\Delta G° = -980.1 + (-1529.078) - (-2639.5) = 130.3\ kJ/mol_{rxn}$

13-57 From the "ideal gas relationship" the pressure is directly related to the mol/L concentration.

$PV = nRT \qquad P = (n/V)RT$ where RT is a constant at a given temperature.

13-59 $K_p = K_c(RT)^{\Delta n}$

$\Delta n = 3 - 2 = 1$

$K_p = K_cRT$, response (c)

13-61

	$COCl_2(g)$	$\rightleftarrows$	$CO(g)$	+	$Cl_2(g)$	$K_p = 3.2 \times 10^{-3}$
initial	0.124		0		0	
change	$-\Delta p$		Δp		Δp	
equilibrium	$0.124-\Delta p$		Δp		Δp	

$$\frac{\Delta p^2}{0.124-\Delta p} = 3.2 \times 10^{-3}$$

Assume $\Delta p << 0.024$

$\Delta p^2 = (0.124)(3.2 \times 10^{-3})$

$\Delta p = 1.99 \times 10^{-2}$

Check, $\frac{1.99\times10^{-2}}{0.124} \times 100\% = 16\%$, the assumption is not valid.

Solve using the quadratic equation.

$$\frac{\Delta p^2}{0.124-\Delta p} = 3.2 \times 10^{-3}$$

$\Delta p^2 = (3.2 \times 10^{-3})(0.124 - \Delta p)$

$\Delta p^2 = 3.96 \times 10^{-4} - 3.2 \times 10^{-3}\Delta p$

$\Delta p^2 + 3.2 \times 10^{-3}\Delta p - 3.96 \times 10^{-4} = 0$

$\Delta p = 1.8 \times 10^{-2}$

$P_{COCl_2} = 0.11$ atm, $P_{CO} = P_{Cl_2} = 1.8 \times 10^{-2}$ atm

13-63

	$2\ SO_3(g)$	$\rightleftarrows$	$O_2(g)$	+	$2\ SO_2(g)$	$K_p = 1.5 \times 10^{-5}$
initial	0.490		0		0	
change	$-2\Delta p$		Δp		$2\Delta p$	
equilibrium	$0.490 - 2\Delta p$		Δp		$2\Delta p$	

$$K_p = \frac{P_{SO_2}{}^2 P_{O_2}}{P_{SO_3}{}^2} = \frac{(2\Delta p)^2 \Delta p}{(0.490 - 2\Delta p)^2} = 1.5 \times 10^{-5}$$

K_p is small and assume Δp is small

$$K_p \approx \frac{4\Delta p^3}{(0.490)^2} = 1.5 \times 10^{-5}$$

$\Delta p^3 = 9.0 \times 10^{-7}$

$\Delta p = 9.7 \times 10^{-3}$

$P_{SO_3} = 0.490 - 2(9.7 \times 10^{-3}) = 0.471$ atm

$P_{SO_2} = 2(9.7 \times 10^{-3}) = 1.9 \times 10^{-2}$ atm

$P_{O_2} = 9.7 \times 10^{-3}$ atm

$$\text{Check, } K_p = \frac{(1.9 \times 10^{-2})^2 (9.7 \times 10^{-3})}{(0.471)^2} = 1.6 \times 10^{-5}$$

13-65 $K_p = \dfrac{P_{NH_3}{}^4 P_{O_2}{}^7}{P_{NO_2}{}^4 P_{H_2O}{}^6} = 1.8 \times 10^{-28}$

$$Q_p = \frac{(0.50)^4(0.50)^7}{(0.50)^4(0.50)^6} = \frac{(0.50)^{11}}{(0.50)^{10}} = 0.5 > K_p$$

The reaction will go to the left.

13-67 In this problem $K_p = K_c$

	$N_2(g)$	+	$O_2(g)$	$\rightleftarrows$	$2\ NO(g)$	$K_p = 4.3 \times 10^{-9}$
initial	0.40		0.60		0	
change	$-\Delta c$		$-\Delta c$		$+2\Delta c$	
equilibrium	$0.40 - \Delta c$		$0.60 - \Delta c$		$2\Delta c$	

$$\frac{(2\Delta c)^2}{(0.40 - \Delta c)\ (0.60 - 3\Delta c)} = 4.3 \times 10^{-9} \text{ at } 700\ ^\circ\text{C}$$

Assume $\Delta c << 0.40$

$$\frac{4\Delta c^2}{(0.40)\ (0.60)} = 4.3 \times 10^{-9} \text{ at } 700\ ^\circ\text{C}$$

$\Delta c = 1.6 \times 10^{-5}$, the assumption is valid.

$[N_2]_{eq} = 0.40$ M

$[O_2]_{eq} = 0.60$ M

$[NO]_{eq} = 3.2 \times 10^{-5}$ M

13-69 The sign of $\Delta G°$ tells you in which direction the reaction will go from the standard state. The magnitude of $\Delta G°$ will tell you how far equilibrium is from the standard state.

13-71 When the reaction quotient, Q, equals one.

13-73 $CO(g) + 2H_2(g) \rightleftarrows CH_3OH(g)$

$\Delta G° = 1040.156 + 2 \times 406.494 - 1877.94 = -24.80$ kJ/mol$_{rxn}$

$$\ln K = \frac{-\Delta G^o}{RT} = \frac{-\left(-24.80 \times 10^3 \text{ J}\right)}{\left(8.314 \frac{\text{J}}{\text{K}}\right)(298.15 \text{ K})} = 10.0$$

$K = e^{10.00} = 2.2 \times 10^4$

13-75 $2HI(g) + Cl_2(g) \rightleftarrows 2HCl(g) + I_2(s)$

$\Delta G° = 2 \times 272.05 + 211.360 - 2 \times 404.226 - 141.00 = -193.99$ kJ/mol$_{rxn}$

$$\ln K = \frac{-\Delta G^o}{RT} = \frac{-\left(-193.99 \times 10^3 \text{ J}\right)}{\left(8.314 \frac{\text{J}}{\text{K}}\right)(298.15 \text{ K})} = 78.26$$

$K = e^{78.26} = 9.7 \times 10^{33}$

13-77 $NH_3(aq) + H_2O(l) \rightleftarrows NH_4^+(aq) + OH^-(aq)$ $K_b = 1.8 \times 10^{-5}$

$\Delta H° = 1205.94 + 970.30 - 1475.81 - 696.81 = 3.62$ kJ/mol$_{rxn}$

$\Delta S° = 386.1 + 320.57 - 498.8 - 286.52 = -78.7$ J/mol$_{rxn}$K

$\Delta G° = \Delta H° - T\Delta S°$

$= 3.62 \times 10^3$ J/mol$_{rxn}$ $- 298.15$ K$(-78.7$ J/mol$_{rxn}$K$)$

$= 2.70 \times 10^4$ J/mol$_{rxn}$

or from standard free energies of atom combination

$\Delta G° = 1091.87 + 875.354 - 1347.93 - 592.222 = 27.07$ kJ/mol$_{rxn}$

$$\ln K = \frac{-\Delta G^o}{RT} = \frac{-\left(2.707 \times 10^4 \text{ J}\right)}{\left(8.314 \frac{\text{J}}{\text{K}}\right)(298.15 \text{ K})} = -10.89$$

$K = e^{-10.89} = 1.9 \times 10^{-5}$ which agrees with the value in the Appendix.

13-79 $CO_2(s) + H_2(g) \rightleftarrows CO(g) + H_2O(g)$

$\Delta H° = 1608.531 + 435.30 - 1076.377 - 926.29 = 41.16$ kJ/mol$_{rxn}$

$\Delta S° = 266.47 + 98.742 - 121.477 - 202.23 = 41.51$ J/mol$_{rxn}$K

$\Delta G° = 1529.078 + 406.494 - 1040.156 - 866.797 = 28.62$ kJ/mol$_{rxn}$

$$\ln K = \frac{-\Delta G^o}{RT} = \frac{-\left(28.62 \times 10^3 \frac{\text{J}}{\text{mol}}\right)}{\left(8.314 \frac{\text{J}}{\text{mol K}}\right)(298.15 \text{ K})} = -11.55$$

$K = e^{-11.55} = 9.6 \times 10^{-6}$. $\Delta H°$ is positive for this reaction and as shown in Chapter 10, K should increase with temperature. $\Delta S°$ is positive for this reaction. Increasing temperature makes $\Delta G°$ more negative, but as shown in the text, the behavior of $\Delta G°$ cannot always be relied upon for predictions about the temperature dependence of K.

13-81 The equilibrium constant for a reaction becomes 1 when $\Delta H° = T\Delta S°$. Assume the standard state enthalpy and entropy changes are approximately those at the temperature desired.

$\Delta H° = 1297.9 - 966.7 - 243.358 = 87.8$ kJ/mol$_{rxn}$

$\Delta S° = 624.60 - 347.01 - 107.330 = 170.26$ J/mol$_{rxn}$K

$$T = \frac{\Delta H°}{\Delta S°} = \frac{87.9 \times 10^3 \text{ J/mol}_{rxn}}{170.26 \text{ J/mol}_{rxn}\text{K}} = 516 \text{ K}$$

13-83 $2SO_3(g) \rightleftarrows 2SO_2(g) + O_2(g)$

$\Delta H° = 2 \times 1422.04 - 2 \times 1073.95 - 498.34 = 197.84$ kJ/mol$_{rxn}$

$\Delta S° = 2 \times 394.23 - 2 \times 241.71 - 116.972 = 188.07$ J/mol$_{rxn}$K

$\Delta G° = \Delta H° - T\Delta S° = 197.79$ kJ/mol$_{rxn}$- T(188.07 J/mol$_{rxn}$K)

$$\ln K = \frac{-\Delta G°}{RT} \text{ and } K = e^{-\frac{\Delta G°}{RT}}$$

T(K)	$\Delta G°$(J/mol$_{rxm}$)	ln K	K
298	1.42×10^3	-57.2	1×10^{-25}
473	1.09×10^3	-27.7	9×10^{-13}
673	71.2×10^3	-12.7	3×10^{-6}
873	33.6×10^3	-4.63	1×10^{-2}

The equilibrium constant increases with increasing temperature. ΔH is positive for this reaction. Therefore, this trend is expected.

13-85 (a) $\Delta S° = 2 \times (188.25) - 2 \times (130684) - (205.138) = -90.01$ J/mol$_{rxn}$K

$\Delta H° = 2 \times (-241.818) - 2 \times (0) - 0 = -483.636$ kJ/mol$_{rxn}$

$\Delta G° = -4.8363 \times 10^5$ J/mol$_{rxn}$ $-$ 298.15 (-90.01 J/mol$_{rxn}$K)= -457.144 kJ/mol$_{rxn}$

$\Delta G° = 2 \times (-228.572) = -457.144$ kJ/mol$_{rxn}$

(b) $\Delta S° = 2 \times (72.13) - 2 \times (51.21) - (223.066) = -181.226$ J/mol$_{rxn}$K

$\Delta H° = 2 \times (-411.1553) - 2 \times (0) - 0 = -822.306$ kJ/mol$_{rxn}$

$\Delta G° = -8.22306 \times 10^5$ J/mol$_{rxn}$ $-$ 298.15 (-181.226 J/mol$_{rxn}$K)= -768.476 kJ/mol$_{rxn}$

$\Delta G° = 2 \times (-384.138) = -768.476$ kJ/mol$_{rxn}$

(c) $\Delta S° = 2 \times (192.45) - 3 \times (130.684) - (191.61) = -198.762$ J/mol$_{rxn}$K

$\Delta H° = 2 \times (-46.11) - 3 \times (0) - 0 = -92.22$ kJ/mol$_{rxn}$

$\Delta G° = -9.222 \times 10^4$ J/mol$_{rxn}$ $-$ 298.15 (-198.762 J/mol$_{rxn}$K)= -32.90 kJ/mol$_{rxn}$

$\Delta G° = 2 \times (-16.45) = -32.90$ kJ/mol$_{rxn}$

In all three cases the $\Delta G°$'s do not change with the method of calculation and are in agreement with the predictions from question 30.

13-87 $NH_3(g) \rightarrow NH_3(l)$

$\Delta H° = -80.29 - (-46.11) = -34$ .18kJ/mol$_{rxn}$

$\Delta S° = 111.3 - 192.45 = -81.15$ J/mol$_{rxn}$K

These values are in agreement with the prediction of problem 31.

13-89 $\Delta G^\circ_{rxn}(HAc) = 2.71 \times 10^4$, $\Delta G^\circ_{rxn}(HCl) = -3.59 \times 10^4$ Computer calculations were used to generate this table and significant figures were ignored.

Qc	ln Qc	ΔGrxn(HAc)	ΔGrxn(HCl)
1.00E-10	-23.026	-29948.204	-92948.204
1.00E-09	-20.723	-24243.383	-87243.383
1.00E-08	-18.421	-18538.563	-81538.563
1.00E-07	-16.118	-12833.742	-75833.742
1.00E-06	-13.816	-7128.9221	-70128.922
1.00E-05	-11.513	-1424.1018	-64424.102
1.00E-04	-9.21	4280.71858	-58719.281
1.00E-03	-6.908	9985.53894	-53014.461
1.00E-02	-4.605	15690.3593	-47309.641
1.00E-01	-2.303	21395.1796	-41604.82
1.00E+00	0	27100	-35900
1.00E+01	2.303	32804.8204	-30195.18
1.00E+02	4.605	38509.6407	-24490.359
1.00E+03	6.908	44214.4611	-18785.539
1.00E+04	9.21	49919.2814	-13080.719
1.00E+05	11.513	55624.1018	-7375.8982
1.00E+06	13.816	61328.9221	-1671.0779
1.00E+07	16.118	67033.7425	4033.74248
1.00E+08	18.421	72738.5628	9738.56283
1.00E+09	20.723	78443.3832	15443.3832
1.00E+10	23.026	84148.2035	21148.2035

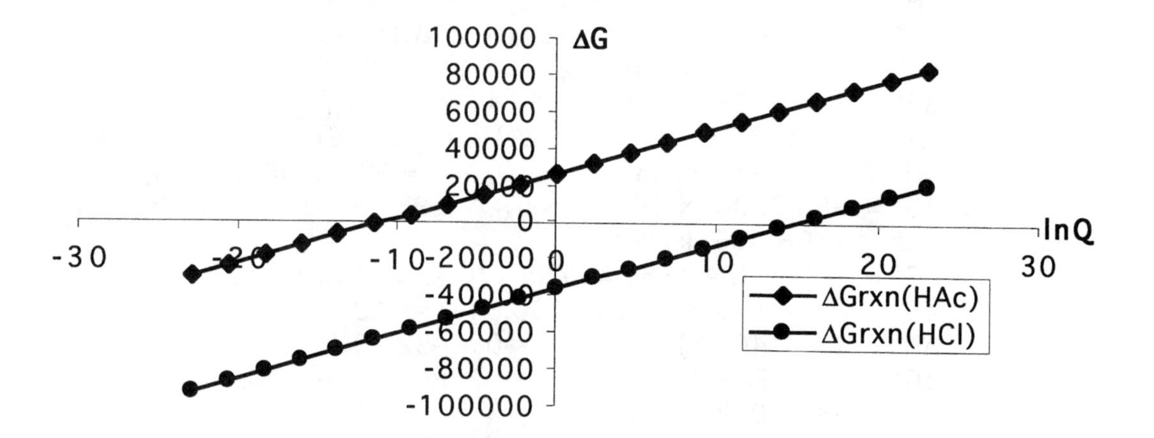

The graph shows that when $\ln Q_c = 0$, $\Delta G_{rxn} = \Delta G^\circ_{rxn}$. When $\Delta G_{rxn} = 0$, the system is at equilibrium, and the point of intersection with the x axis is the ln K for the reaction.

13-91 $2\ NO(g) + O_2(g) \rightleftarrows 2\ NO_2(g)$

(a) $\Delta H^\circ = 2 \times (-937.86) - 2 \times (-631.62) - (-498.340) = -114.74\ kJ/mol_{rxn}$

(b) $\Delta S^\circ = 2 \times (-235.35) - 2 \times (-103.592) - (-116.972) = -146.54\ J/mol_{rxn}\ K$

(c) $\Delta G^\circ = -\ 114.74\ kJ/mol_{rxn} - 298(-146.54\ J/mol_{rxn}K) = -\ 71.1\ kJ/mol_{rxn}$

(d) At 298K the Gibbs energy for the reaction in the standard state is negative so it is a favorable reaction.

(e) $$\ln K = \frac{-\Delta G^\circ}{RT} = \frac{-\left(27.14 \times 10^3\ J\right)}{\left(8.314\frac{J}{K}\right)(298.15\ K)} = -10.95$$

$K = e^{28.7} = 2.86 \times 10^{12}$.

(f) $$K_p = \frac{P_{H_2}{}^3 P_{N_2}}{P_{NH_3}{}^2} = \frac{(6.0)^3(20)}{(0.018)^2} = 1.3 \times 10^6.$$ Since $Q < K$ the reaction will proceed to the right.

(g) If temperature is decreased the entropy term in ΔG° will decrease in magnitude, making the ΔG° more negative; this will increase the equilibrium constant.

13-93 Using the data from 13-91 and examining 13-91c in particular. Increasing the temperature will decrease the magnitude of ΔG°.

13-95 $H_2O(aq) \rightleftarrows H^+(aq) + OH^-(aq)$

$\Delta H^\circ = -218 + (-696) - (-970) = 56\ kJ/mol_{rxn}$

$\Delta S^\circ = -115 + (-286) - (-321) = 56 = -80\ J/mol_{rxn}\ K$

$\Delta G^\circ = 56\ kJ/mol_{rxn} - 298(-80\ J/mol_{rxn}K) = 80\ kJ/mol_{rxn}$

$CH_3COOH(aq) \rightleftarrows H^+(aq) + CH_3COO^-(aq)$

$\Delta H^\circ = -218 + (-3071) - (-3288) = -1\ kJ/mol_{rxn}$

$\Delta S^\circ = -115 + (-896) - (-919) = 56 = -92\ J/mol_{rxn}\ K$

$\Delta G^\circ = -1\ kJ/mol_{rxn} - 298(-92\ J/mol_{rxn}K) = 26\ kJ/mol_{rxn}$

Acetic acid will be stronger. Even though both reactions have a positive ΔG°. The magnitude of the ΔG° for the first reaction is much greater, so it will be less likely to proceed than the second reaction.

Chapter 14
Kinetics

Note to student: The data plots given in the solutions to these problems were easily generated using Microsoft Excel©. The fitted lines were created by adding a trendline to the plot. It is a worthwhile exercise for you to learn how to generate these on your own.

14-1 **Thermodynamic control** of reaction takes place when a reaction takes place the way the thermodynamics predicts it should. That is, addition of magnesium metal to a strongly acidic solution will produce Mg^{2+} ions and hydrogen gas because the thermodynamics say that's what should happen. However, sometimes a reaction which is thermodynamically favored is kept from proceeding rapidly because it is under **kinetic control**. Such a reaction is the combustion of octane to produce carbon dioxide and water. While this is a thermodynamically favored reaction, it does not happen instantly when oxygen and octane are present in an automobile engine.

14-3 Since it is thermodynamically favorable the reaction can reasonably be expected to proceed. However, there is a kinetic barrier to the reaction which keeps it from happening rapidly at room temperature and one atmosphere.

14-5 The rate is expressed in the form of a change in one of the reactants or products as a function of change in time.

14-7 As the reaction proceeds forward the phenolphthalein is consumed in the reaction, so as the concentration decreases (a negative change) we must show the rate as being positive (moving forward), therefore we put a negative sign in front of the change in concentration of a reactant.

14-9 The instantaneous rate is given by rate $= -\frac{\Delta x}{\Delta t}$. Which is simply the slope of the curve of the concentration (x) at a given time (t).

14-11 For most reactions, the rate of reaction changes with time. The instantaneous rate of reaction is for a very short period of time. It can be measured more accurately than the changing rate of reaction, over a long period.

14-13 If the amount of N_2O_4 is measured in mol/liter, the units for the rate constant are time^{-1}.

14-15 If the amounts are measured in mol/liter, the units for the constant are M^{-2} time $^{-1}$.

14-17 The larger bore indicates a faster rate, which means that the rate constant must be larger for the larger bore. (X) represents the volume of water passing through. The units of k in this system are sec^{-1}.

14-19 For each Cl_2 consumed, $2ClF_3$ are formed, therefore the rate of appearance of ClF_3 is twice the rate of depletion of Cl_2. Response (c).

14-21 The reaction stoichiometry indicates that the rate of disappearance of HI is twice as great as the rate of formation of I_2 .

$$-\frac{d(HI)}{dt} = 2\frac{d(I_2)}{dt} \qquad \frac{d(I_2)}{dt} = -\frac{1}{2}\frac{d(HI)}{dt} = -0.5x(-0.039\ M/s) = 0.020M/s$$

14-23 $$\frac{d(NO)}{dt} = \frac{d(H_2O)}{dt} x \frac{4molNO}{6molH_2O} = 0.040\frac{M}{s} x \frac{4}{6} = 0.027\ M\ s^{-1}.$$

$$-\frac{d(NH_3)}{dt} = \frac{d(H_2O)}{dt} x \frac{4molNO}{6molH_2O} = 0.027\ M\ s^{-1}$$

14-25 The molecularity is the same as the stoichiometry and describes the ratio of the substances involved in a chemical reaction and the relative amounts of each in the change described. The order of the reaction describes the dependence of the rate upon the concentrations of the reactants.

14-27 Most reactions having multiple mechanistic steps leading from reactants to products are likely to have rate laws which differ from what the stoichiometry of the reaction would lead us to expect. The rate limiting step may fall early in the sequence of steps of the mechanism.

14-29 Rate = $k \cdot (A) \cdot (B)$

14-31 A reaction **mechanism** is the sequence of simple chemical steps which take place in the sequence of a reaction.

14-33 The identification of second order reactions suggests that the associated elementary reactions that occur in a single step might be sought among those with two reactants and no more. Thus, we reject responses (a) and (c) for having more than two reactants. Response (b) displays a single reactant. Only response (d) remains.

14-35 The second order rate equation suggests a slow, rate limiting step involving $NO_2 + F_2$, followed by a fast step for incorporation of the additional NO_2. Response (c).

14-37 From the slow step we determine,
Rate = $k(I^-)(HOCl)$
HOCl does not appear in the overall equation. The concentration of HOCl is related to the concentration of OCl^- through the equilibrium constant for the first step in the reaction mechanism.

$$K_{eq} = \frac{[HOCl][OH^-]}{[OCl^-]} \text{ from which } (HOCl) = \frac{K_{eq}(OCl^-)}{(OH^-)}$$

Substituting into the rate expression, $\text{Rate} = k\,K_{eq}\frac{(OCl^-)(I^-)}{(OH^-)}$

NO, the mechanism is not consistent with the rate law.

14-39 From the slow step: Rate = $k\,(Br)(H_2)$

From step 1: $K_{eq} = \dfrac{[Br]^2}{[Br_2]}$

$[Br]^2 = K_{eq}\,[Br_2]$

$[Br] = K_{eq}^{1/2}[Br_2]^{1/2}$

Substituting into the rate equation, Rate = $k\,K_{eq}^{1/2}(Br_2)^{1/2}(H_2)$

14-41 $Keq = \dfrac{k_f}{k_r} = \dfrac{5.2x10^{-7}M^{-1}s^{-1}}{1.5x10^{-11}M^{-1}s^{-1}} = 3.5x10^4$

14-43 A reaction can be zero order when the reaction rate is not dependent upon the concentrations of any of the reactants, but may be held up, for example, by the ability of those reactants in finding a site on a catalyst.

14-45 Rate = $k\,(NO)^x(Cl_2)^y$

From the data for the first two reactions, doubling the concentration of NO causes the initial rate of reaction to increase by a factor of 4. The value of x in the reaction equation is 2. From the data for the third and fourth reactions, doubling the concentration of Cl_2 causes the rate of the reaction to also double. The value of y in the reaction equation is 1.

Rate = $k\,(NO)^2(Cl_2)$

14-47 If the volume is constant, the partial pressure of a compound is directly proportional to its concentration.

Rate = $k\,(NO)^x(O_2)^y$

From the data for the first two reactions we find that $(100/150)^x = 0.355/0.800$. The reaction is second order in NO (x = 2). From the data for the last two reactions (130/180) = (1.04/1.44). The reaction is first order in oxygen. Rate = $k\,(NO)^2\,(O_2)$.

14-49 From the initial reaction rate data for the first and second reactions we find that 1.35/1.00 = 8.78/6.50. The reaction is first order in iodomethane. x = 1. From the data for the last two reactions, 0.15/0.25 = 8.29/13.8. The reaction is first order in hydroxide ion. y = 1.

Rate = $k\,(CH_3I)\,(OH^-)$

14-51 The derivative form of the rate equation is best used to work with problems concerned with how fast a reaction proceeds under a given set of concentrations. The integrated form of the rate equation is used to determine the value of the rate constant. Given the rate constant, the integrated form of the rate equation can be used to give information on the amounts of reactants or products at a given time.

14-53 Rate = k = $3.4 \times 10^{-6}\ M^{-1}s^{-1}$. After 1000 sec, $k \cdot t$ of the reactant(X_o=0.0068 M) will be used up.

$X-X_o = -k \cdot t = -3.4 \times 10^{-6}\ M^{-1}s^{-1} \cdot 1000\ sec = 0.0034\ M$. So X=0.0034 M left.

14-55 (a) $\ln \frac{X}{X_o} = -kt$

$\frac{\ln(0.50)}{-k} = t$

$\frac{\ln(0.50)}{-5.6x10^{-2}s^{-1}} = t = 12\ s$

(b) $\ln \frac{X}{X_o} = -kt$

$\frac{\ln(0.50)}{-k} = -5.6 \times 10^{-2}\ s^{-1}$

X = 0.49 M

14-57 For a first order reaction $t_{1/2} = \frac{0.693}{k}$

$k = t_{1/2} \frac{0.693}{t_{1/2}} = \frac{0.693}{0.0282s} = 24.6\ s^{-1}$

14-59 For second order

$\frac{1}{X} - \frac{1}{X_o} = kt,\ \frac{1}{X} - \frac{1}{0.55M} = 9.6\times10^{-2}\,L\cdot mol^{-1}\cdot min^{-1}\times10\,min$, X= 0.36 M.

14-61 $\ln \frac{X}{X_o} = -kt$

$\ln \frac{(Rb)}{(Rb)_o} = (-1.42 \times 10^{-11}y^{-1})(1.19 \times 10^{10}\ y) = -0.169$

$\frac{(Rb)}{(Rb)_o} = e^{-0.169} = 0.845$, 84.5%

14-63 $\ln \frac{X}{X_o} = -kt,\ t_{1/2} = \frac{0.693}{k},\ k = \frac{0.693}{t_{1/2}}$

$\ln \frac{X}{X_o} = \frac{-0.693}{5730y}(15520y) = -1.88$

$\frac{X}{X_o} = \frac{^{14}C}{^{14}C_o} = e^{-1.877} = 0.15$, 15%

14-65 $\frac{\ln(0.623)}{-0.693/5730y} = 3.91 \times 10^3 y$

14-67 The detection limit is 0.10 in 15.3 so $\frac{x}{x_o} = \frac{0.10}{15.3}$

$$\frac{\ln\frac{0.10}{15.3}}{-0.693/5730y} = 4.2 \times 10^4 \text{ y}$$

14-69 In comparing the first concentration data to the second to the last we find that the half-life is 480 s, $k = \frac{0.693}{480} = 1.44 \times 10^{-3} s^{-1}$

$\ln \frac{(N_2O)}{(N_2O)_o} = - kt$
$\ln (N_2O) - \ln (N_2O)_o = -kt$
$\ln (N_2O) - \ln (0.100) = - 1.44 \times 10^{-3} s^{-1} (900 \text{ s})$
$\ln (N_2O) - (- 2.30) = - 1.35$
$\ln (N_2O) = - 3.65$
$(N_2O) = 0.026$ M

14-71 The slope of the line in problem 14-70 is the rate constant.

$$- k = \frac{\ln 0.100 - \ln 0.046}{(0 - 144)h} = - \frac{-2.302 - (-3.08)}{-144}$$
$= 5.4 \times 10^{-3} h^{-1}$

$k = 5.4 \times 10^{-3} h^{-1}$

$$t_{1/2} = \frac{0.693}{k} = \frac{0.693}{5.4 \times 10^{-3} \ h^{-1}} = 1.3 \times 10^2 \text{ h}$$

14-73 If the rate of reaction does not depend on the concentration of PPh_3, being first-order in $Ni(CO)_4$, the rate equation for this reaction is consistent with the mechanism as given. Rate=$k(Ni(CO)_4)$

14-75 Since the time for one half of the original sample to react is 1000 sec and the time for the next half of the reactants to be removed is 2000 sec, this means this is a second order reaction.

14-77 The partial pressure of a gas is proportional to the concentration of that gas (see problem 14-47).

A plot of ln of the pressure of dimethyl ether vs. time gives a linear plot of slope = -k. $k=0.0129\ min^{-1}$. So $t_{1/2} = \frac{0.693}{k} = \frac{0.693}{5.4 \times 10^{-3}\ h^{-1}}$. This matches the time it takes for the initial pressure to decrease by one half.

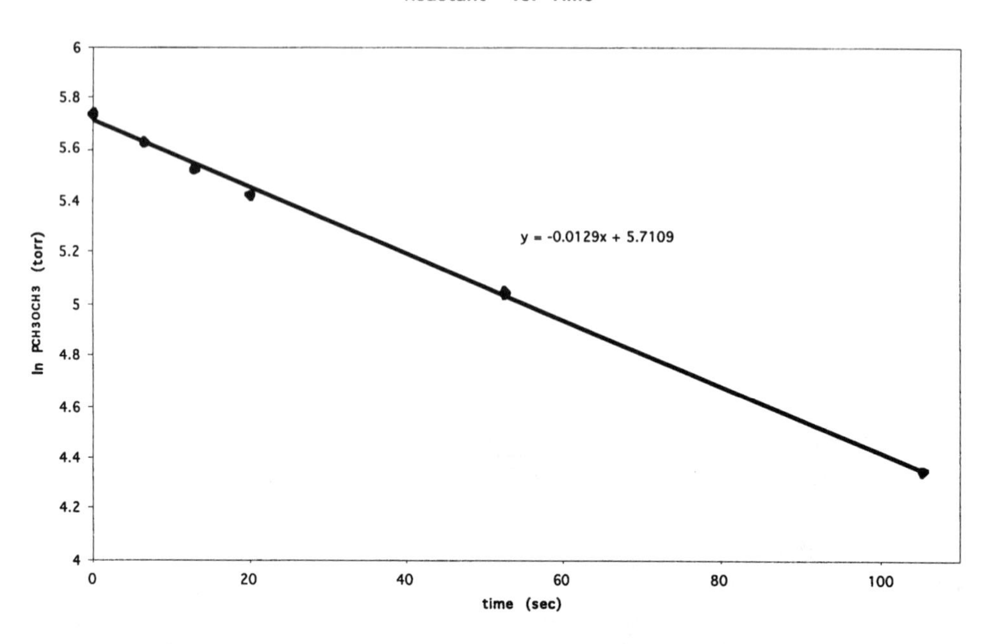

14-79 A plot of the ln $(Cr(NH_3)_5Cl^{2+})$ versus time gives a most nearly straight line, indicating a possible first-order reaction. The plot of $1/(Cr(NH_3)_5Cl^{2+})$ is not as linear. The reaction is most likely first-order in $(Cr(NH_3)_5Cl^{2+})$.

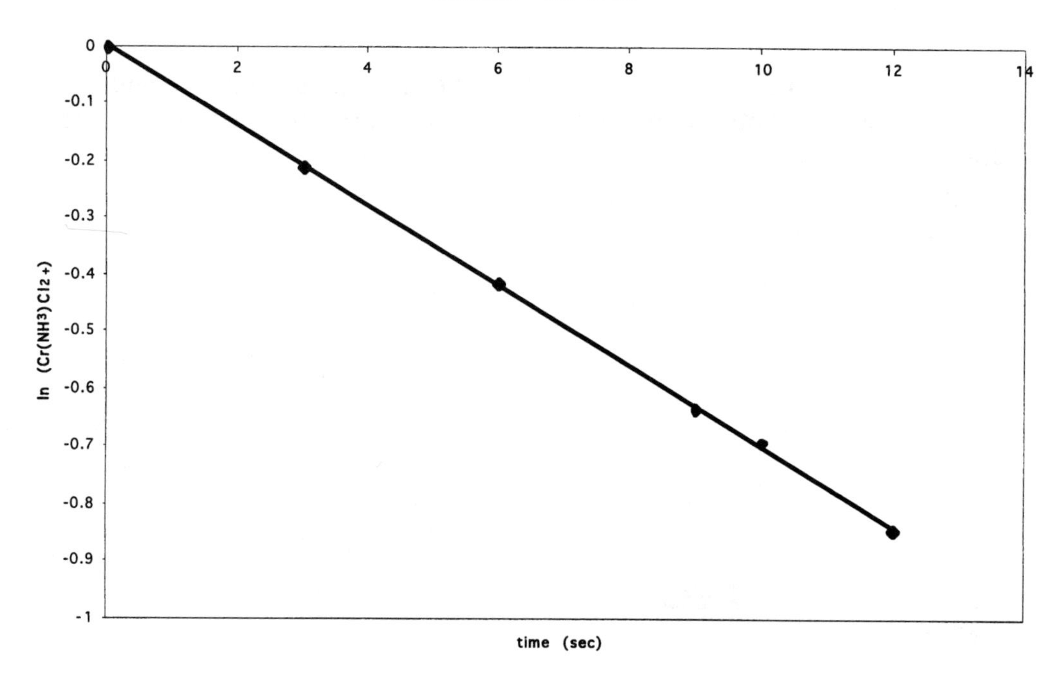

14-81 The slow step of the given mechanism is bimolecular, first-order in each of the reactants. In the presence of an excess of OH^-, the reaction becomes pseudo-first-order in the metal ion complex. None of the subsequent steps of the mechanism will affect the observed reaction rate. The rate equation for the mechanism described is

$$\text{rate} = k\,(Cr(NH_3)_5Cl^{2+})(OH^-)$$

14-83 (d) In order to make it pseudo-first-order the OH^- concentration should be as big as possible, and definitely at least two orders of magnitude greater than that of CH_3I.

14-85 Collision theory says that the higher the temperature the more collisions will take place between molecules in a given time period and the higher the energy of the collision. Both of these factors increase the likelihood that in a given time frame more successful reactions will occur.

14-87 For simple molecule reactions, the relative energy of the collisions between reacting molecules determines whether the collision may be effective and lead to products, or ineffective with the interacting molecules returning to reactants. For some reactions, the relative orientation of the colliding particles will have an influence.

14-89 Lowering the activation barrier will increase the rate of the reaction, since more molecules will have enough energy to get over the barrier.

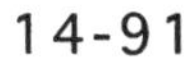
14-91

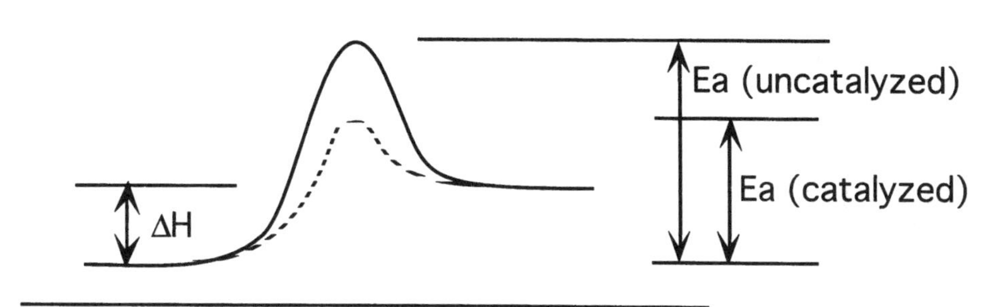

Reaction coordinate

14-93 Assume that the activation energy remains essentially constant for small changes in temperature.

$$k = Z\,e^{-Ea/RT}$$

The ratio of the rate constants is

$$\frac{k1}{k2} = \frac{Z \cdot e^{-Ea/RT_1}}{Z \cdot e^{-Ea/RT_2}} = e^{-E_a/RT_1}\,e^{+E_aRT_2} = e^{\frac{Ea}{R}\left(\frac{1}{T2}-\frac{1}{T1}\right)}$$

$$\frac{k1}{k2} = \frac{1.52\text{x}10^{-5}\,s^{-1}}{3.83\text{x}10^{-3}\,s^{-1}} = 3.97 \times 10^{-3}$$

$$\ln 3.97 \times 10^{-3} = \frac{E_a}{8.314}\left(\frac{1}{318}-\frac{1}{298}\right) = E_a\left(-2.54 \times 10^{-5}\right)J$$

$$E_a = \frac{-5.529}{-2.54 \times 10^{-5}} = 2.18 \times 10^5\,J_{mol_{rxn}} = 2.18 \times 10^2\,kJ_{mol_{rxn}}$$

14-95 The plot of ln k versus 1/T gives a straight line with a slope of -4.02×10^3. The Arrhenius equation can be written,

$$\ln k = \ln Z - \frac{Ea}{RT} \text{ where the slope} = \frac{-Ea}{R}$$

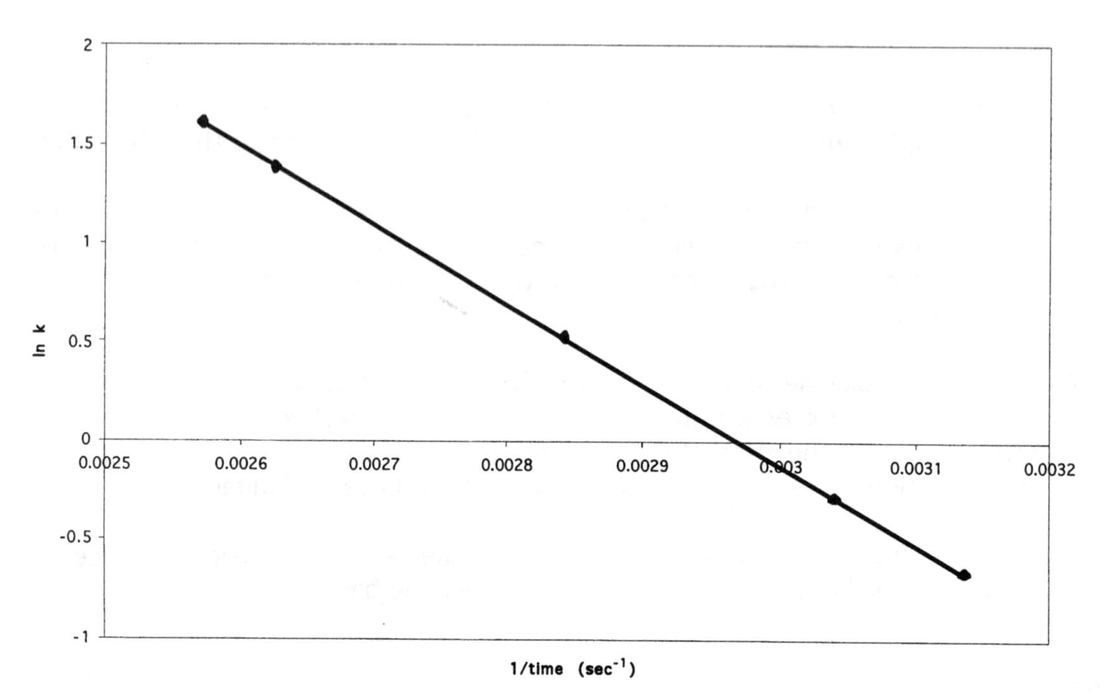

$$Ea = -(-4.02 \times 10^3 \text{ K})\times(8.314 \frac{J}{mol \cdot K}) = 33.4 \frac{kJ}{mol_{rxn}}$$

14-97 Assume E_a does not change significantly with small changes in temperature.

$$\ln \frac{k1}{k2} = \frac{Ea}{R}\left(\frac{1}{T_2} - \frac{1}{T_1}\right)$$

$$\ln (6.5 \times 10^{-5}) - \ln k_2 = \frac{92.9\text{x}10^3 \frac{J}{mol_{rxn}}}{8.314 \frac{J}{mol_{rxn} K}}\left(\frac{1}{348} - \frac{1}{298}\right)$$

$k_2 = 0.014 \text{ M}^{-1}\text{s}^{-1}$

14-99 At low concentrations of sucrose, the enzyme converts sucrose to products. If a little more sucrose were added, the reaction rate would increase, and experiment has shown that the reaction is first-order at low concentrations of sucrose.

V(ml)	ln V	time(s)
50	3.9	0
40	3.7	19
30	3.4	42
20	3.0	72
10	2.3	116
0		203

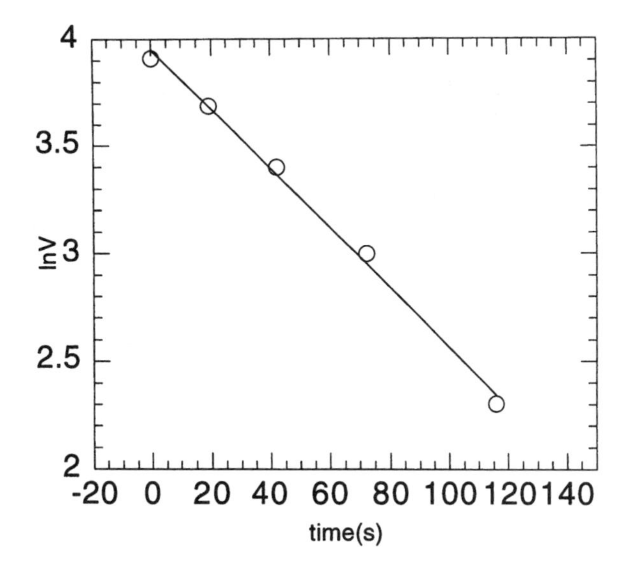

14-101 A plot of ln V versus time gives a nearly straight line indicating a first-order process. The rate constant is the negative of the slope of the line.

$$\text{slope} = -k = \frac{3.9 - 2.3}{0 - 116}$$

$= -1.4 \times 10^{-2}$ ml s^{-1}

14-103 A plot of volume versus time produces a straight line. The process is zero-order. A plot of lnV vs time produces a curved line.

14-105 (a) IV, (b) I, (c) III, (d) III,
Since the products are more stable than the reactants in II, this reaction would have K>1. II and IV appear to have similar activation energies and would proceed at comparable rates.

14-107 If we assume that reaction must occur by the coming together of the reactants, the rate of "coming together" must be directly related to the amount of reactants available to come together. The amount of reactants available to come together is measured by the concentrations of the reactants consumed in that step.

14-109 As the temperature of a reaction mixture increases, the number of collisions between particles having sufficient energy to cause the transformation to products increases. The description of the reactivity of hydrogen and oxygen closely follows this pattern. At lower temperatures, the reaction experiences kinetic control.

14-111 (a) Rate = -k (H_2O_2), because the first step is the rate determining step. The reaction is therefore first-order with respect to (H_2O_2).

(b) The data also indicate first-order with respect to (H_2O_2). The $t_{1/2}$ = 835 sec. After 1670 sec (2 x $t_{1/2}$) only one fourth of the reactant is present. This matches the relationship for a first-order reaction given in Table 14.4.

(c) The $t_{1/2}$ = 835 sec. See (b).

(d) The activation energy in the reverse reaction is larger since the starting point will be lower in energy for the products than for the reactants.

(e)The turnip (somehow) is acting as a catalyst for the reaction. By increasing the rate by 600 x 10^9. The Ea will be reduced by ln (600 x 10^9) or a factor of 27.

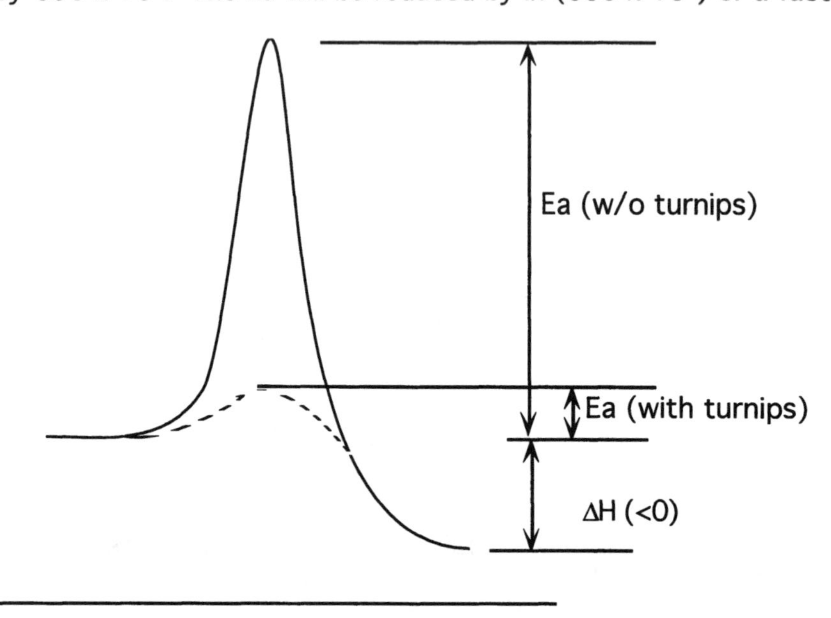

14-113 Mechanism (c) is consistent with this rate law since the slow step (step 2) requires one H_2 and one N_2O_2 , the latter being formed in the previous reaction which requires two NO molecules. Mechanism (b) is also consistent with the rate law since the stoichiometric coefficients of the slow step are the orders of the reactants in the rate law.

14-115 (a) Rate = k $(NO)^2(Cl_2)$. This is because doubling the (Cl_2) doubles the rate, while doubling the [NO] quadruples the rate.

(b) 0.18 M/sec = k $(.10\ M)^2(0.10\ M)$, k= 180 $M^{-2}\ sec^{-1}$.

(c) The mechanism is consistent since the slow step requires a molecule of NO and $NOCl_2$, the latter being formed from a second NO and a Cl_2.

14-117 (a) This reaction is second-order with respect to A because the time for half the reactants to react from 1.000 M to 0.500 M is half that from 0.500 M to 0.250 M (see Table 14.4).

(b) Rate = k $(A)^2$.

(c) $t_{1/2} = \dfrac{1}{k(1.000\,M)} = 40\,sec$, k = 0.025 $M^{-1}\ sec^{-1}$.

(d) The catalyst will increase the rate constant because it would increase the overall rate of the reaction.

14-119 The data appears to suggest a second-order reaction, because the time for one half consumption appears to increase as the reaction proceeds, plotting the reciprocal of concentration vs. time gives a linear plot. The rate constant is the slope = 0.0821 M^{-1} sec^{-1}.

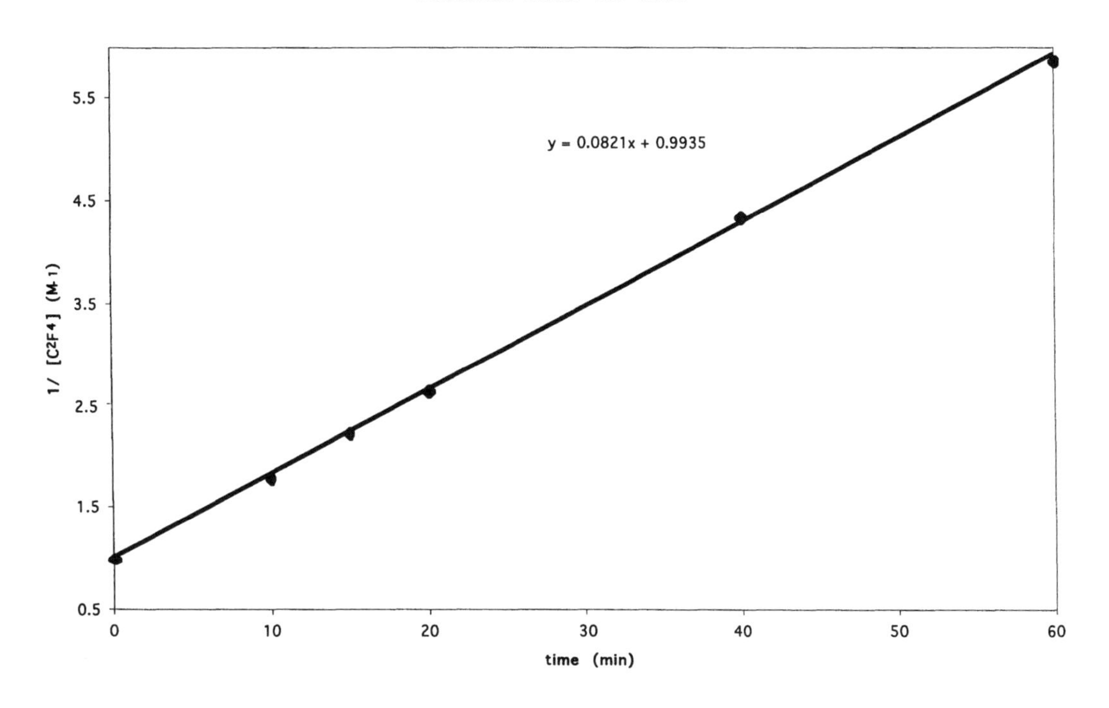

Chapter 15
Chemical Analysis

15-1 In a **quantitative** analysis, the amount of analyte is determined. In a **qualitative** analysis, the identity of the substance is determined. In a **structural** analysis, the structure of the analyte is determined.

15-3 A mixture is separated by chromatography by the interaction of the components of a mixture with a mobile and stationary phase. The mobile phase pushes the components through a column while the stationary phase retards the movement of the components. The differing degrees of interaction of the mixture components with the mobile and stationary phases lead to the separation of components.

15-5 The interactions of the mixture components with the mobile and stationary phases are often based on intermolecular forces of attraction.

15-7 (b) and (e). These two compounds are very structurally similar and thus have very similar retention times.

15-9 A mass spectrum is a plot of relative intensity (detector response) versus mass/charge ratio of the fragment ions striking the detector. A mass spectrum can be used for the determination of molecular weight or the identification of chemical substances.

15-11 (b) heptane; the molecular weight of heptane is 100 g/mole. The molecular ion peak in the spectrum is assigned a mass of 100.

15-13 $A = \varepsilon bc$ where A=absorbance, ε=molar absorptivity, b=path length, and c=concentration.

15-15 Each technique uses a different type of electromagnetic radiation. This causes different changes within the molecule and yields different information about the molecule. NMR gives different signals based upon the chemical environment the various nuclei are in. IR examines molecules by looking at the different kinds of bonds. Each bond has a different vibrational frequency. UV/Vis typically gives only a few broad features based on the electronic transitions of the molecule.

15-17 (a) absorbance increases; (b) absorbance decreases; (c) absorbance decreases

15-19 There are four peaks in the spectrum indicating that there are four kinds of protons present. This could be (a) or (b). However, the single and separate peak at 9.8 ppm is indicative of very different chemical environment, such as the CHO group. This indicates that the compound is butyraldehyde.

15-21 For the first experiment the answer is (c) the sample is reacting and is being used up so absorbance decreases until there is no more sample remaining.

For the second experiment the answer is (a) the concentration of product increases and therefore absorbance increases until there is no more product produced.

15-23 0.37

15-25 Beer's law states that absorbance is directly proportional to concentration. The concentration of X will decrease with respect to time. This suggests that either b or d is correct. Chapter 14 described that for a first-order reaction, the decrease in concentration of a reactant will follow a logarithmic curve. Plot d corresponds to a first-order reaction. Plot b will correspond to the zero-order reaction.

15-27 Current instrumentation allows detection of pesticides at such small concentrations that almost all food will show some measurable amount of pesticide due to residual pesticides in soil from many years ago or from airborne sources.

15-29 Ginger ale is a mixture; therefore, the first step is to separate it into its components using a chromatographic method. HPLC is the technique of choice. A chromatogram is also run of a normal ginger ale sample and the two chromatograms compared. If a component exists in the suspect ginger ale that is not in the normal sample, it should be isolated from the sample and identified using IR, NMR, or mass spectroscopy.

15-31 The NMR spectra for the two compounds would be different. Both compounds have six different types of hydrogens although the CH_2 groups are so similar their peaks would probably overlap. The peak ratios for the first compound would theoretically be:

3:2:2:2:2:1 as compared to the second compound: 3:1:1:2:2:3. This would give a distinct difference in these two spectra.

15-33 (a) Since the stationary phase is nonpolar, polar molecules will separate first. Molecule A is the most polar because of a C=O bond and an OH group. Molecule C is the next most polar and Molecule B is the least polar. The order is A then C then B. The figure on the lower right matches this order.
(b) Peak area is proportional to relative amounts. Therefore, A and C are present in the same amount. Molecule B's concentration is one and a third greater.
(c) Yes, Molecule A would show strong C=O and OH absorbances. Molecule B would show only C-C and C-H interactions. Molecule C would show distinguishable C-O stretches.